T0354149

The Sausage of the Future

A research project by Carolien Niebling conducted at ECAL
Published by Lars Müller Publishers

‘A highbrow is someone who looks at a sausage and thinks of Picasso.’

— A. P. Herbert

Preface

According to current thinking around the sustainability of our planet, we should aim to reduce meat consumption and increase dietary diversity. For this project, I brought together a chef of molecular gastronomy, a master butcher and a designer to look into sausage production techniques and potential new ingredients with which to create a 'future sausage' that would challenge the traditional design of this food item and seek to improve and adapt it to the realities that await us. Among the many publications consulted during the making of this book, Harold McGee's *On Food and Cooking* and Nathan Myhrvold, Chris Young and Maxime Bilet's *Modernist Cuisine* were major sources of inspiration, and provided the basis for several sections of this book.

Whenever I mention my research on the sausage of the future, people tell me enthusiastically about their favourite type of sausage, or mention a fond memory associated with it. Most often, the sausage discussed is specific to their culture, and is appreciated mostly because of its historical, cultural and personal associations, sometimes even over its actual flavour.

Throughout history, humans have shown great inventiveness in the face of sustenance scarcity. Hence the sausage, one of mankind's greatest edible inventions – and one of the first processed (or designed, if you will) food products – arose from the need to use every scrap available.

A sausage is not just a source of nourishment. It is an object designed to provide protein on long journeys and to extend the shelf life of meat, but also to combine certain ingredients (scraps) and disguise others (offal). Yet these functions are compromised in our contemporary food culture, for the ease with which we obtain our groceries from supermarkets allows us to create a distance between ourselves and the origins of our food. This distance has made us pickier in our food choices, and the industry's handling and disguising of our food has created a hypocrisy under which we are allowed to grow ever more detached from, even disgusted by, what we consume.

The aim of this book is to provide readers with enough information to understand the complexity of sausages and to see their beauty.

I wish to empower readers to change their food habits and to imagine the redesigning of sausages using existing techniques, rather than relying on lab-based innovation. Instead of having mushroom-based vegetarian options that taste like chicken, or milk serum-based edible packaging, should we instead make use of readily available flowers, neglected offal and undervalued grains?

I propose here a move away from our relatively impoverished 'supermarket selection', a move back from what I call *eximius forivores* (eaters of supermarket food) to *omnivores* (eaters of everything). Let us embrace what is edible on this planet and explore it in the name of diversity.

Contents

Preface 5

The Anatomy of the Sausage 9

Theory 17

History of the Sausage & The Future of Food 19

The Meat we Eat 25
Beef, meat fibre, pork and muscle colour

Skins of all Sorts 35
Intestines, collagen, cellulose, nylon, fabric, fibre-reinforced, wax, algae, leaves and gelatine

Method 39

Preservation & Glue 40

Production Techniques 45
Mincing, filling & extruding and linking

Types of Sausage 51
Fermented, fresh, dried, cooked and blood

Material 59

Proteins & Protein Chart 61

From Eximius Forivore to Omnivore 67
Offal, insects, plants, seeds, grains, nuts and legumes

Result 109

The Art of Mixing & Sausage Matrix 111

Futurivore 117
Mortadella with vegetables, bangers and mash & apricot and carrot & pea and chickpea, heart fuet, fruit salami, liver and berry, apple boudin and Insect pâté

The Psychology of Disgust and the Desire for Delicacy 147

Glossary 149

Sausage Matrix 150
Bibliography 153
Biographies 155
Colophon 156

The elements of a sausage

- Mass
- Moisture
- Flavour
- Salt
- Skin

The Anatomy of the Sausage

Sausage *noun*

A mixture of minced edible substances encased in an edible or non-edible skin to form a new food item. Whether or not it contains meat, if it is encased in a skin it could be considered sausage.

From Latin *salsus*, salted.

What constitutes a sausage differs widely from country to country. In the Czech Republic, a *salami* would never be called a sausage, and in Britain, *brawn* (also known as head cheese) is considered a meat preparation and surely not a sausage. In this book, we present a broad view of the definition of a sausage, which makes the possibilities greater and the results more surprising. The sausage is easily deconstructed into a few key components: mass, moisture, glue, flavour and skin.

Mass makes up the biggest part of the sausage. At around 65 per cent, it might also be referred to as bulk, body, corpus, lion's share or meat. Mass can be made of one or a combination of ingredients such as meat, fat, offal, grains, legumes, seeds, fruit, vegetables, and so on. It is often a dense substance, containing little water. Its purpose is to be the body of the sausage and to hold its shape; it also needs to be sympathetic to the 'glue', or to become so during cooking.

Moisture is the second biggest part of the sausage. Originally the mass is made of a combination of lean offcuts and trimmings: pieces that are tough naturally. Therefore, traditionally around 30 per cent of the sausage is water and fat. The latter gives – besides moisture and tenderness – a great deal of flavour, while the added water content makes for better mixing. In general, added moisture is divided into two categories: water-based and fat-based. However, the basic rule is for the combined mass and moisture never to contain more than 15 per cent of water, unless it is for a gelatinous type of sausage such as head cheese.

Glue is what holds together all the ingredients of a sausage. It seldom constitutes a large percentage of the contents of the sausage, but it is crucial to its integrity. The glue, or gel, is also referred to as the binding of the sausage. How you bind the sausage depends on what ingredients you are using and how you treat them. Often ingredients already contain a natural binder that just needs to be activated, for example by being combined with other ingredients, heated up or broken down into little particles that form new bonds.

Flavouring is only a small percentage of the sausage, but one which has a huge impact on the final product. Besides the usual salt and pepper, the most common spices are mace, cloves, garlic, onion, nutmeg, cinnamon, sage and rosemary – a collection of highly aromatic ingredients. Salt is not there simply for its taste but also for its flavour-enhancing properties, as it lifts the aromatic tones of the other ingredients.

Preservation is what makes the ingredients of a sausage stay fresh, by creating an environment in which bad microbes cannot multiply. These environments include a pH-level of under 4, a high salt level, lack of oxygen or water, exposure to smoke, heat, cold and good bacteria. It can even be the addition of certain ingredients such as beetroot, rosemary or cranberry. It is possible to combine one or more of these ingredients or processes to achieve a satisfactory end result.

> *The environment has a big impact on the flavour of a sausage, but even more so on its texture.*

Skin has an obvious primary task and should work closely together with the preservative(s). A sausage with the wrong skin can spoil by the effect of the preservative. For example, a dried sausage requires perishable skin in order to release the moisture. Or, when a sausage is to be poached in water, the skin needs to protect the mass from mixing with the water. Originally, when sausages were the result of efficient butchery, there was an abundance of casings available in the animal naturally. However, since around the middle of the 20th century, there have been many different types of casings invented that go beyond the intestines. These are created to serve new purposes, such as suitability for vegetarians, or reinforcement for a more mass-scale production.

Collage deconstructing a mortadella on a background of a macro image of broccoli, alongside the traditional pork, also broccoli, carrots, romanesco, cauliflower and pistachio nuts.

Collage deconstructing an insect paté on a background of a macro image of a grasshopper wing, alongside variations on traditional ingredients: ewe's milk foam, mealworms, Madeira wine and butter.

Collage deconstructing a heart fuet on a background of a macro image of beef heart, alongside some of the fuet's pre-minced ingredients: fennel seeds and salt crystals.

Collage deconstructing a bangers and mash sausage on a background of a macro image of chia seeds, alongside some of its pre-minced ingredients: chia seeds, potato and peas.

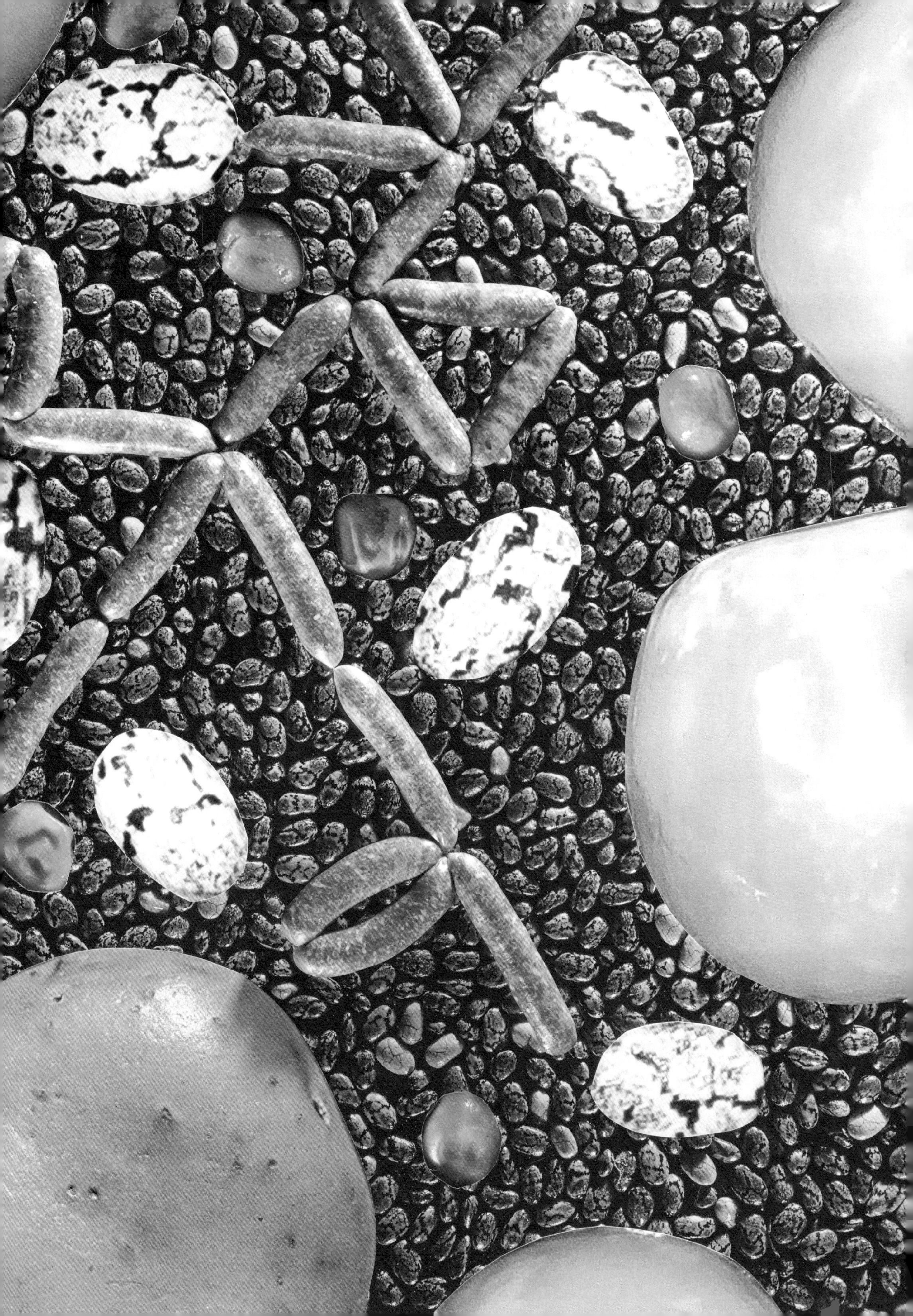

Theory

History of the Sausage & The Future of Food

The following essay lays out the basic background information needed to understand why we need to look into the design of the sausage and use this as a medium to eating differently.

25 The Meat we Eat
27 Beef
29 Meat fibre
31 Pork
33 Muscle colour
35 Skin

At the Ter Weele butchery, Netherlands: a collection of artisanal sausages, soon to be given extra depth in flavour by smoking.

History of the Sausage

The earliest method of food preservation used by mankind was drying. The place in which it was discovered, now called the Sahara Desert, is one of the hottest and driest areas on the planet. Food history constitutes the foundation for a better understanding of sustenance: why we eat certain things and how. Back when we still had to hunt animals to consume them, no scrap of meat was wasted, as hunting was a physically demanding and unpredictable source of food. In order to use all the edible parts as well as the convenient natural vessels and casings available, the butcher would put the offal, scraps, ears, skin and fat into the stomach and bowels of the animal to create a piece of food in an edible casing. Hence the sausage was born.

The first records of the existence of sausages date from the Bronze Age (3300 BCE – 2000 BCE) in the Middle East. It was at around the same time that the ancient Greeks discovered that salting meat prolonged its life. It therefore became traditional for butchers to salt various tissues such as scraps, organ meats and fat to help preserve the meat. (The word *sausage* is derived from the Old French *saussiche* and from the Latin *salsus*, both meaning 'salted'.) The use of the tube-like intestines led to the invention of the characteristic cylindrical-shape of the sausage as we know it. Hence, sausages, puddings and salami are among the oldest prepared foods in the world. Food historians believe that the Romans picked up the craft from the Lucanians, a tribe that for almost 1,000 years ruled part of what is now Basilicata in southern Italy, developing a reputation for sausages while fending off imperial conquerors. The popularity of sausages was evident: the Greek poet Homer mentioned a kind of blood sausage in *The Odyssey*, Epicharmus wrote a comedy titled *The Sausage*, and Aristophanes' play *The Knights* is about a sausage vendor who is elected leader. In ancient Rome, there was a thriving meat industry that produced sausages similar to the ones we still enjoy today, such as bratwurst, boudin and fuet. But there were also sausages that have been lost over time, such as brain sausages, which contained brain matter, egg, pine nuts, fish stock, wolf's milk, and spices.

In different parts of Europe, characteristic sausages arose owing to climatic variances. For example, dried sausages originated from naturally warm and sunny climates, whereas cured, fermented or smoked sausages sprang from colder and more humid climates. Furthermore, non-meat additions such as fruit, grains or vegetables were regional choices that depended on local availability. This is the reason why the name of a sausage often indicates its origin: *Frankfurter, Bologna, Lincolnshire,* etc. From the Middle Ages onwards, new sausage recipes were being invented rapidly. Crusaders brought back new herbs and spices from their travels, and soon sausages contained spices such as nutmeg, mace, pepper, sage and cloves – none of which is traditionally European.

At the end of the 19th century, Charles Feltman, a butcher of German origin, introduced the *dachshund* sausage to New York. The sausage quickly became popular because it was a cheap and easy snack. It wasn't until years later that Feltman paired up with his baker brother-in-law to design the matching bun, creating what we now know as the iconic *hot dog*.

Today, the sausage remains a cornerstone of our food culture. Britain alone has over 470 different types of breakfast sausage, and in Germany there are even sausage laws, dictating specific rules for the making of sausages. There are also many sausage associations, such as the *AAAAA, L'Association Amicale des Amateurs d'Andouillette Authentique* (The Friendly Association of Amateur Makers of Authentic Andouillette), a club formed by several food writers in France in the 1970s. The AAAAA gives certificates to producers of high-quality andouillettes, a rustic sausage made from tripe and pork, usually eaten hot, with mustard. The andouillette has a pungent and distinctive aroma. Sausage recipes have not changed much since the mid-20th century, and although the necessity of making sausages became close to irrelevant with the invention of the refrigerator, the craft of sausage making seems unlikely to become obsolete, simply because the product is too versatile, unique and enjoyable.

World population to 2050
(b = billion)

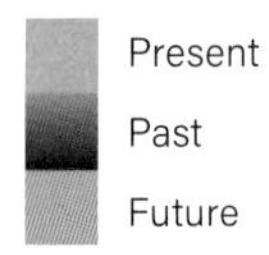

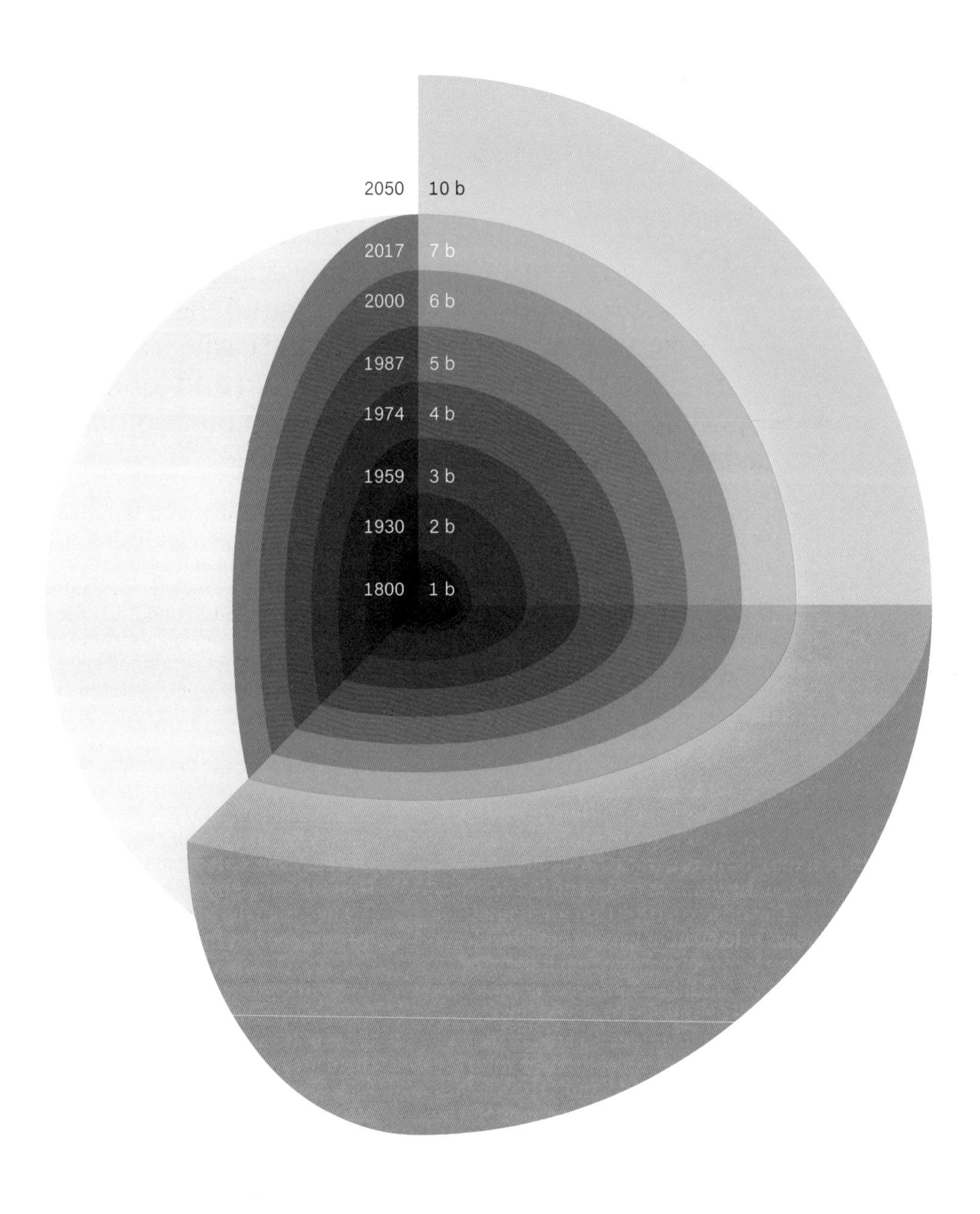

The world's population is growing exponentially. The orange section illustrates an increase of 3 billion worldwide from 2017 to 2054.

The Future of Food

Meat has always been the most highly praised of all the foods. To uncover the source of this prestige, we must look to human nature. Until 2 million years ago, our primate ancestors lived almost exclusively on plant foods. Then, the changing African climate caused vegetation to diminish, leading our ancestors to look towards animal carcasses for their food. Indeed, animal flesh and the fat contained within bone marrow were highly concentrated sources of energy and tissue-building protein. They literally helped to feed the human brain, exploding its growth and thus marking the evolution of early hominids into humans. Later, by virtue of its protein content and by remaining edible for years, dried meat made it possible for humans to migrate from Africa towards Europe and Asia, and thus thrive in cold regions where plant foods were scarce or even absent during long winters. Humans became active hunters around 100,000 years ago, and in the cave paintings they left us depicting wild cattle and horses, it is clear that these animals were painted as embodiments of strength and vitality. By extension, these qualities began to be associated with meat as well, and a successful hunt has long been a proud and grateful occasion for a celebratory feast. We no longer need to hunt our meat, or indeed depend on it for survival; however, animal flesh remains the centrepiece of meals the world over.

The population of the world has increased enormously since the start of the industrial revolution, which began in the 1780s, but it exploded just after the discovery of penicillin in 1928. This growth is tied to three major factors: as well as penicillin, other medicines and vaccines were discovered, thus helping to cure people or preventing them from contracting deadly bacteria and diseases. Moreover, public health improved rapidly with the arrival of running water and sanitation. Finally, the industrialization of agriculture and transportation has made food production and distribution cheaper and more efficient, as well as increasing product availability. Thus, from around the second half of the 20th century, consumers would come to demand whatever food they wanted, whenever they wanted it; seasonal fruits and vegetables were therefore made available year-round by importing them from afar or growing them in protected and suitable environments. Ever since, the food industry has grown exponentially to meet the demands of consumers. But in a world with a swiftly growing population, the prognosis of food supply is cause for concern. For example, the worldwide consumption of meat has doubled in the last twenty years, and is expected to double again by 2050. Therefore, our current appetite for meat demands an intensive and highly efficient method of animal rearing. To make cows grow faster, their normal food of grass or hay (both of which are indigestible for humans) is replaced with grain, thereby requir-ing the industry to use a great deal of land and water, which evidently has a substantial environmental impact. By the end of the 21st century, there will simply not be enough land to produce this much meat for 10 billion people. Furthermore, the use of grains for cattle is an enormous waste of potential food, since grains are suitable for human consumption: to produce 1 kilogrammes of beef, 7 kilogrammes of grain are needed.

Should we all become vegetarians?

To put it simply, meat is an immensely rich source of protein. No longer eating meat as part of our diets would mean substituting this lack of protein with higher quantities of food (because they naturally contain less protein). Alternatively, if the land now used for meat rearing were used for plant crops instead, all the nutrients from the soil

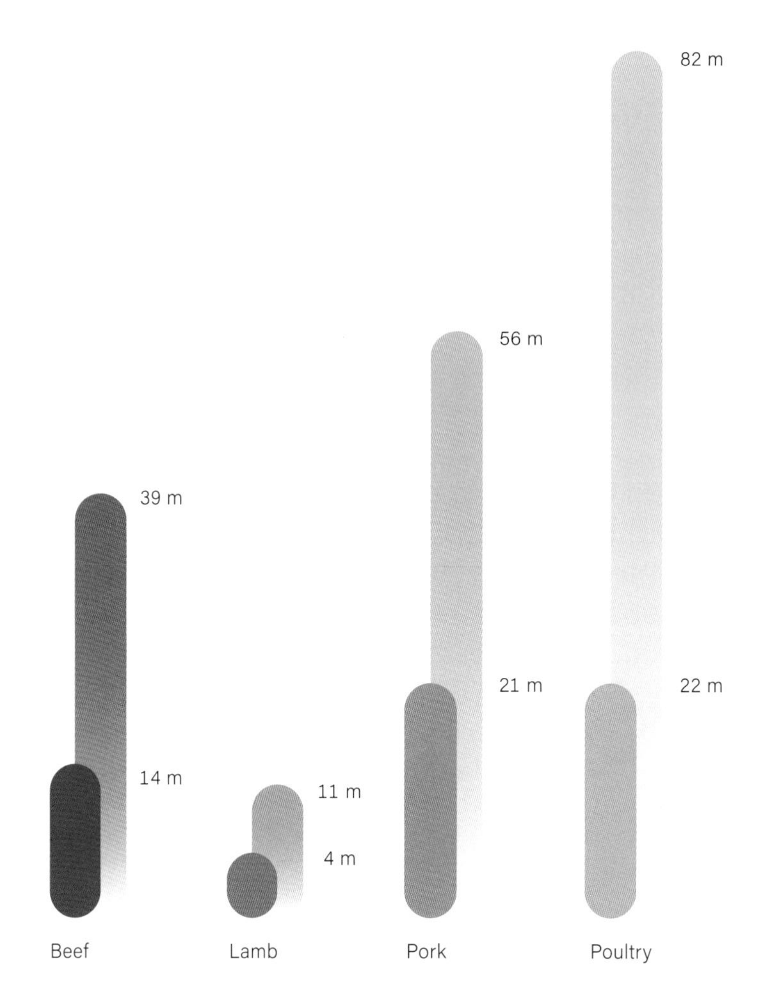

would be used up to the point where it became too poor for crops to grow on. There is therefore no need to eliminate meat form our diets entirely; it is merely time to reduce the quantity that we consume currently to slow down the pace of animal production.

Time for innovation!

There are over one million different types of insect on the planet, and 1,900 of these are considered edible. Further, there are serious advantages to rearing insects for human consumption. For one, insect farming does not require land clearing to increase production. On top of that, insects emit considerably fewer greenhouse gases (GHGs) than most livestock (methane, for instance, is produced by only a few insect groups, such as termites and cockroaches). And because insects are cold-blooded, they are efficient at converting feed into protein. Crickets, for example, need twelve times less feed than cattle, four times less feed than sheep, and half as much feed as pigs and chickens to produce the same amount of protein. Most edible insects are very high in protein, and contain nutrients such as calcium, zinc and iron. The official term for insect eating is *Entomophagy* and currently around 80 per cent of the world's inhabitants are entomophagists, so why aren't we? According to current research by the FAO, Food and Agriculture Organization and the University of Wageningen, there is no question that over the next few decades we will begin eating insects in the West. The only problem is that introducing a completely new type of food into current systems of production and governance will take time. When insects are dead they decompose quickly, because of their high protein content, so insects either have to be alive, frozen or freeze-dried to be ready for human consumption, which is why it is important to have local farms. To create these farms, certain state rules must first be adapted, since it is currently forbidden to grow insects for human consumption in many European countries. Then, developing these farms can take an additional several years, especially if they are to be efficient enough to make the insects affordable. In the meantime, we are still facing a scarcity of protein.

There are many more protein sources being developed rapidly, such as the invention of healthy plant-based alternatives to eggs, chicken,

Meat efficiency chart: percentages represent the quantity edible from the animal's weight when alive.

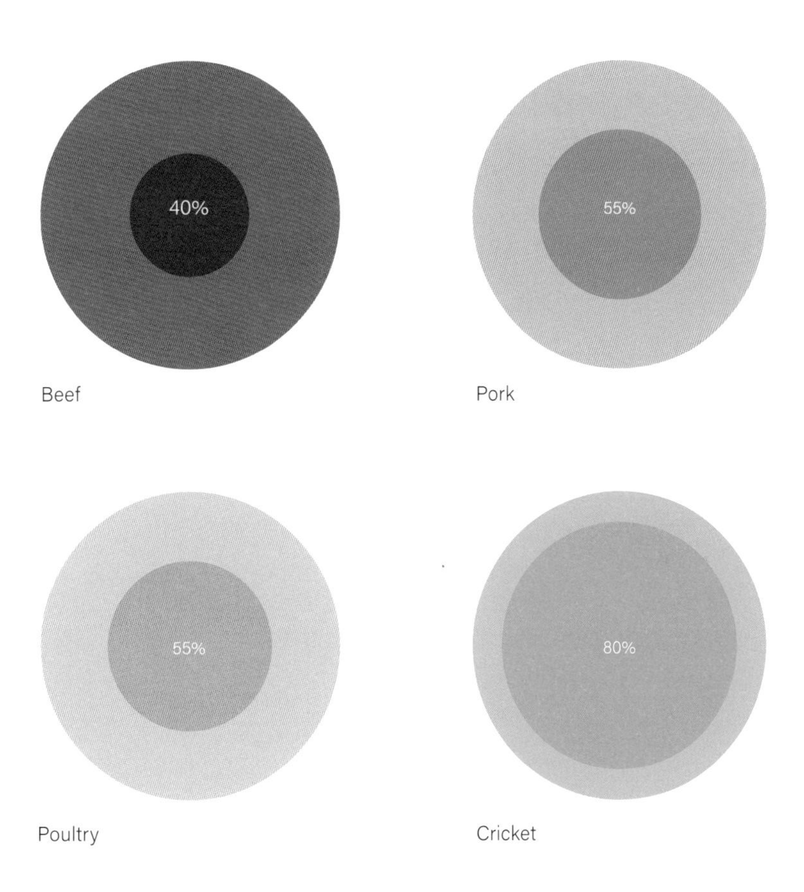

and other sources of protein. Companies such as Beyond Meat and Hampton Creek Foods are experimenting with new ways to use heat and pressure to turn plants into foods that look and taste just like meat and eggs. Moreover, some researchers expect that *in vitro* meat, grown from stem cells in a bioreactor, could provide a sustainable and animal-friendly alternative to conventional meat. In 2013, the first lab-grown burger was cooked and tested, but given that the cost of its production remains at 125,000 euros, it is not yet a viable alternative.

All these solutions maintain the amount of meat we consume, whether it is real, fake or lab-grown. Unfortunately, the production of fake meat only creates a bigger gap between consumers and their food. The idea that we could be eating real or fake chicken without being able to tell the difference is alarming, and it certainly does not encourage us to examine our eating habits more closely and therefore make changes for the better.

Since there are so many of us on this planet, we should use it to our advantage: before these new proteins are feasible, making small adjustments can have an enormous impact. Cutting back on our meat consumption by 10 to 20 per cent would have major consequences in terms of meat production. For example: had we consumed 1 per cent less meat per capita in 2010 (that is, if the average person ate 40 kilogrammes of meat rather than their usual 40.5 kilogrammes), the total consumption of 280,000 million kilogrammes would have been reduced by 3.460 million globally, which is the equivalent of about 5.3 million cows.

The sausage emerged from food scarcity and waste efficiency. So now, when it is yet again crucial to be efficient but also inventive with our food, should we turn to the sausage once more?

At the Ter Weele butchery, Netherlands: ribs on a beef carcass.

The Meat we Eat

Farming animals for human consumption started around 7000 BCE, when people in the Middle East managed to tame a handful of wild animals – first dogs, then goats and sheep, then pigs and cattle and horses – to live alongside them. Livestock not only transformed inedible grass and scraps into nutritious meat, but constituted a walking larder, a store of concentrated nourishment that could be harvested whenever it was needed. When it was time to slaughter an animal, the family or community came together to help process the meat and since inevitably there are some parts that need to be eaten quite quickly afterwards, such as kidneys, the event rapidly became a festive occasion. Even now, in some regions in Italy, France and Eastern Europe, it is a major yearly event where everybody gathers together and processes the meat coming from the animal. Now, the event is often related to the hunting season.

Before meat is meat, it is a muscle. When an animal is alive its muscles are converting chemical energy into mechanical activities such as flying, swimming or running. In the kitchen we can still find that many of these activities have influenced the meat: its texture, colour and flavour. When cooking meat the primary goal is to keep its tenderness. There is something strangely pleasing and unique to the texture of meat, a certain resistance which we don't find in many other foods and if we do we call it 'meaty'– as, for example, mushrooms which can be described as 'meaty', though there is a very fine line between pleasantly resistant and shoe leather tough.

The existence of fat and the density of muscle fibre and connective tissue all contribute to a piece of meat being perceived as tough; a feeling that arises when one chews for a prolonged and somewhat disagreeable period. This texture tends to correlate with the part of the mobility levels of an animal, whether it is young or old, and the location on the animal's body from which a cut of meat was taken. Thus, whereas pasture animals do not use their backs much (think of the little used back muscle aptly called *tenderloin*) other body areas are often utilised, for example, the chest, front legs, neck and shoulders. Accordingly, chicken breasts are built mostly of light muscle fibres because they are used very sporadically to avoid hungry preditors and not constantly to, for example, fly long distances. Therefore the connective tissue in the breast is relatively weak and lean, with a mere 2 per cent of collagen. The legs and thighs, on the other hand, need more stamina to stand, walk and run. So they contain more dark endurance fibres, which burn fat for fuel and are reinforced by collagen (5–8 per cent). This makes it necessary to cook the leg meat longer than the breast meat, but the breast meat can be easily overcooked, whereas the leg meat is more likely to stay succulent and tender.

We all have favourite cuts of meat and a preference for certain animals, but do we know the body parts that we appreciate so much?

The fact that we do not rear the animals whose meet we consume nor prepare our own meat makes us more dependent on the industry to do it for us, which means that this knowledge is rapidly being lost. The contents of the sausage have shifted slowly from scraps to sometimes the best cuts of meat. In Germany, there are several sausage-related laws that determine which particular part of the animal goes in which sausage, and those parts tend to be the prime cuts. But I would like to propose that the sausage should be used for what it was designed to do: containing ingredients that are available and affordable, as they become minced anyhow. The following chapter gives a short animal anatomy lesson – for those who have forgotten or were never taught.

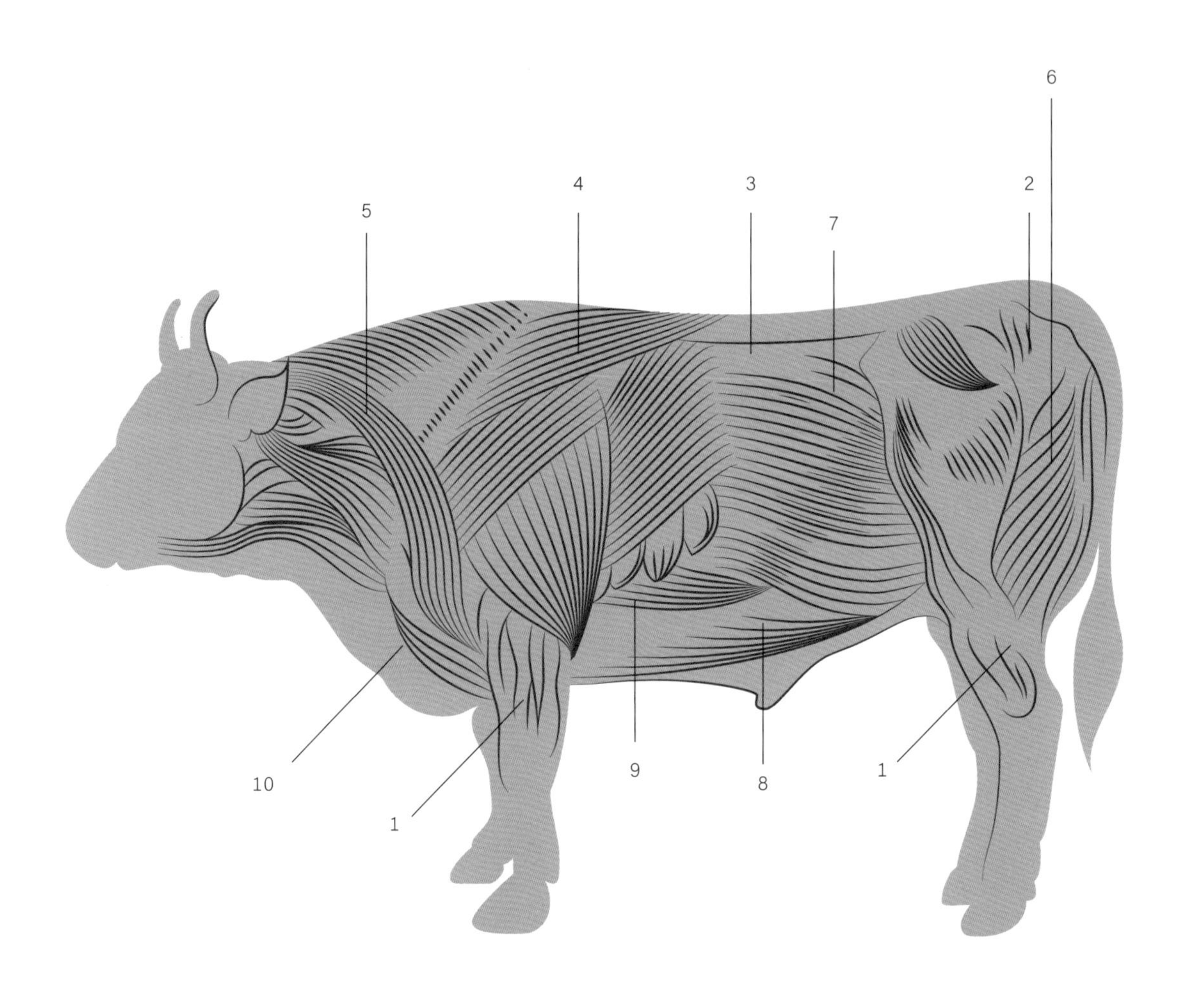

Above and opposite page, top: diagrams giving an overview of general North American cuts and muscle direction. However, different butchers will have their own variation on cutting meat.
Opposite page, bottom: composition of a cow.

Beef

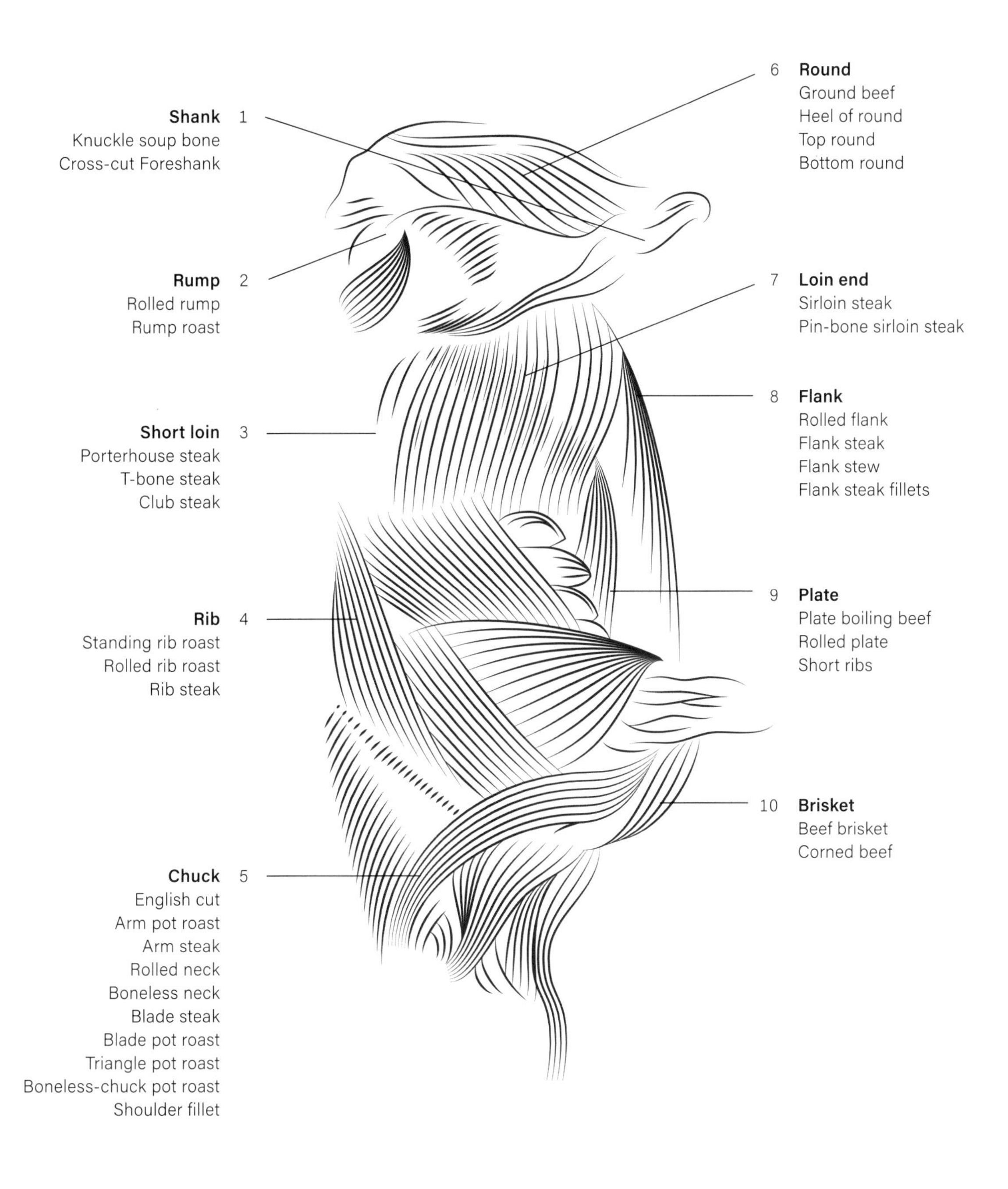

53% Flesh and fat
18% Bones
8% Blood
8% Skin
13% Intestines

Average weight of a cow: 550 kg

The grain of every muscle comes from individual bundles of muscle fibres. Each bundle is known as a fascicle. In general, the finer the grain, the more tender the meat.

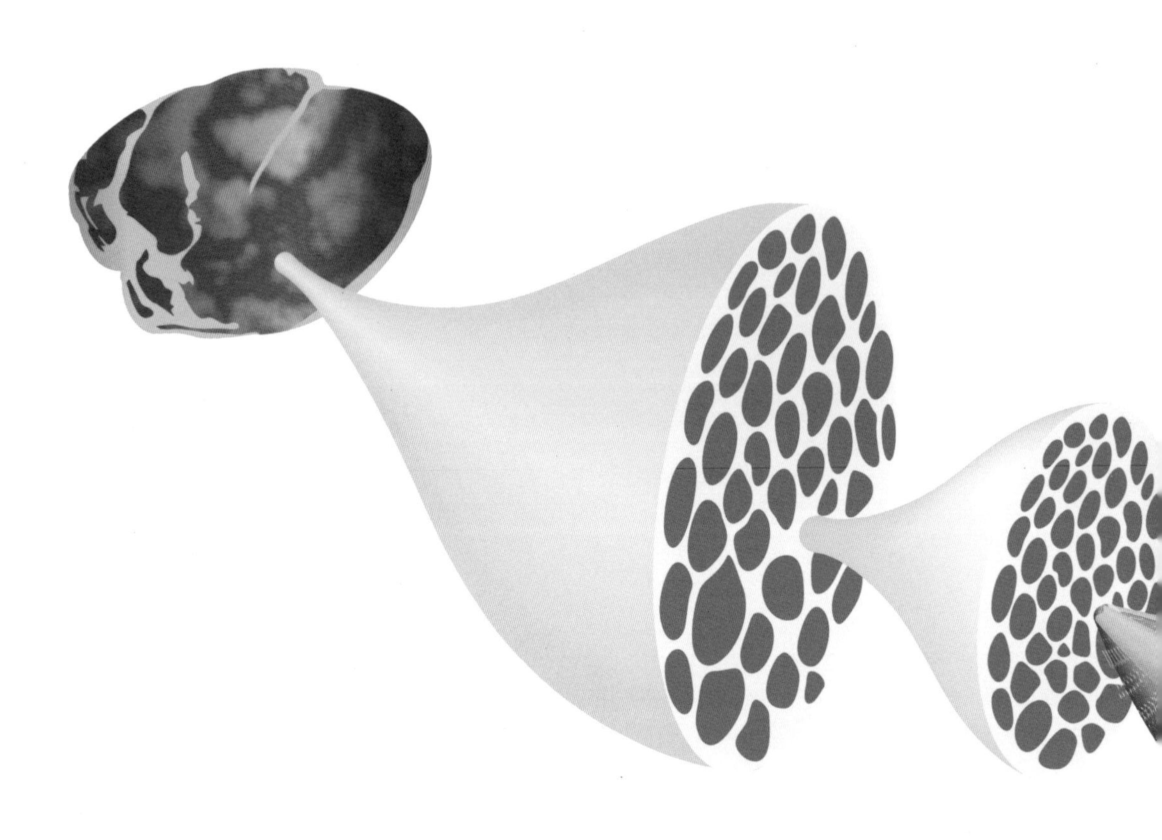

Texture of meat is determined by its structure. Lean meat is low in fat – it stands for 3 per cent of its make-up – high in water content – accounting for 75 per cent of its composition – and contains 20 per cent protein. This trio of major elements combine to form three tissue types. Muscle cells form the main tissue, and their movement is created by tensing and slackening these elongated fibres. The second type saddles together the bones and muscle cells, encompassing the fibres; it is known as connective tissue. The third is located in clusters between the connective tissue and the fibres, and takes the form of fat cells. These clusters operate as stores of energy. The arrangement and relative proportions of these three components will determine the qualities of the meat – its texture, colour and flavour. Bundles of muscle cells – the ones responsible for movement – are visible to the eye, and while one fibre can be the length of an entire muscle its thickness is roughly that of a human hair. When eating meat that has been cooked thoroughly, the muscle fibres can be cut and seen without effort.

Fat enhances the tenderness of meat in three ways: its cells weaken the sheet of connective tissue and the mass of muscle fibres; it melts when heated (instead of drying out and stiffening as the fibres do); and it lubricates the tissue. Tender meat would be compacted, dry and tough without fat. For instance, beef shoulder muscles contain more connective tissue than the leg muscles, but they also

Meat fibre

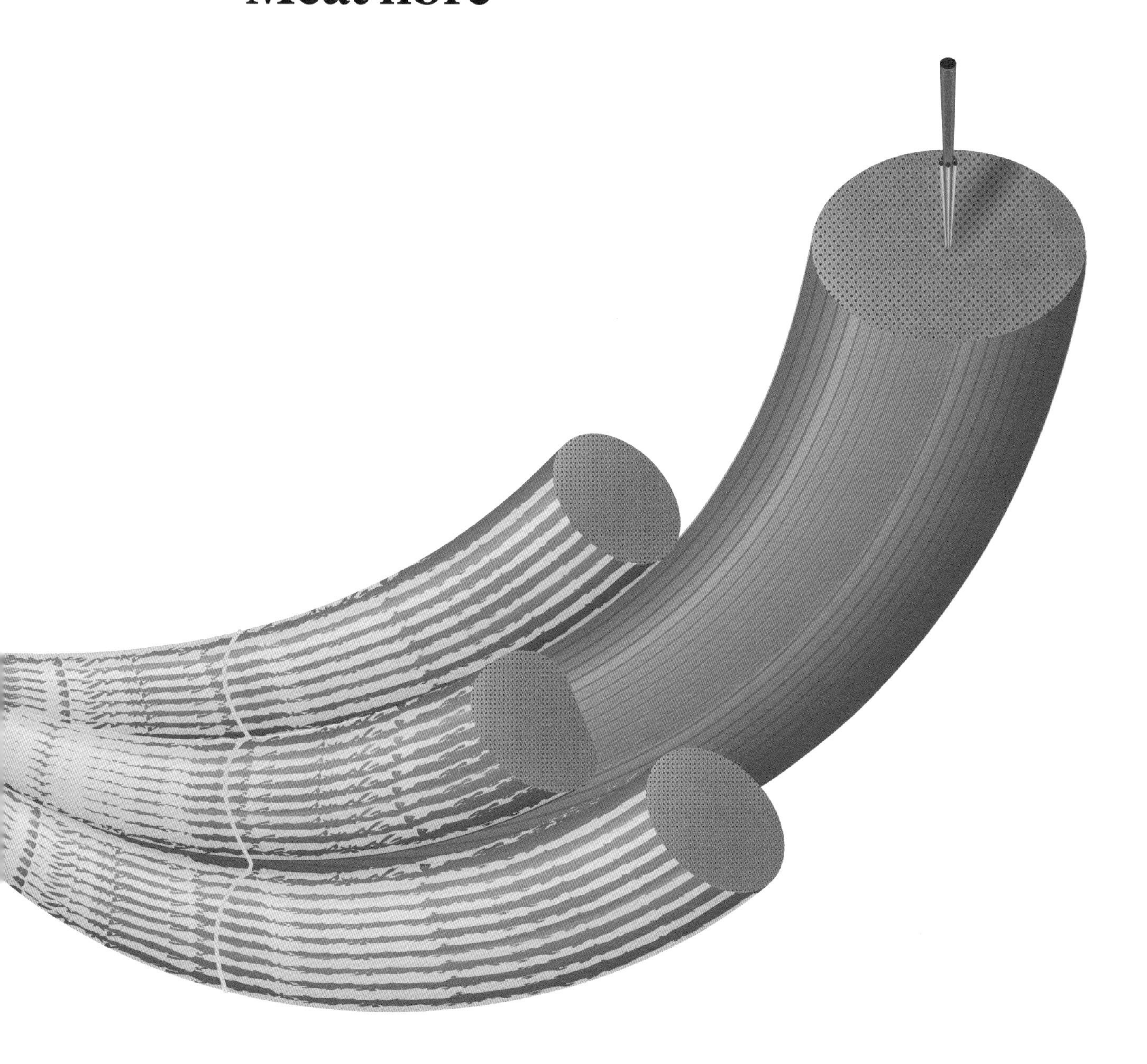

include more fat, and therefore make more succulent dishes. The major filament in connective tissue is a protein called *collagen*, which is concentrated in skin, tendons and bones. The name comes from the Greek for 'glue producing', because solid, tough collagen partly dissolves into sticky gelatine when heated in water. Thus, during cooking, the connective tissue becomes softer, unlike the muscle fibres, which become tougher.

The texture of tender meat is as distinctive and satisfying as its flavour: a 'meaty' food is something you can sink your teeth into, dense and substantial, initially resistant to the tooth but soon giving way as it liberates its flavour.

The dense and firm texture of meat as we know it comes from the mass of muscle fibres, which become denser, dryer and tougher when cooked. Their elongated arrangement makes up the 'grain' of meat. If you cut parallel to the bundles, you will see them from the side, all lined up; if you cut across the bundles, you will just see their ends. It is easier to dislodge fibre bundles from each other than to break the bundles themselves, so naturally it is easier to chew along the direction of the fibres than across them. That is why we usually carve meat across the grain, so that we can chew with the grain.

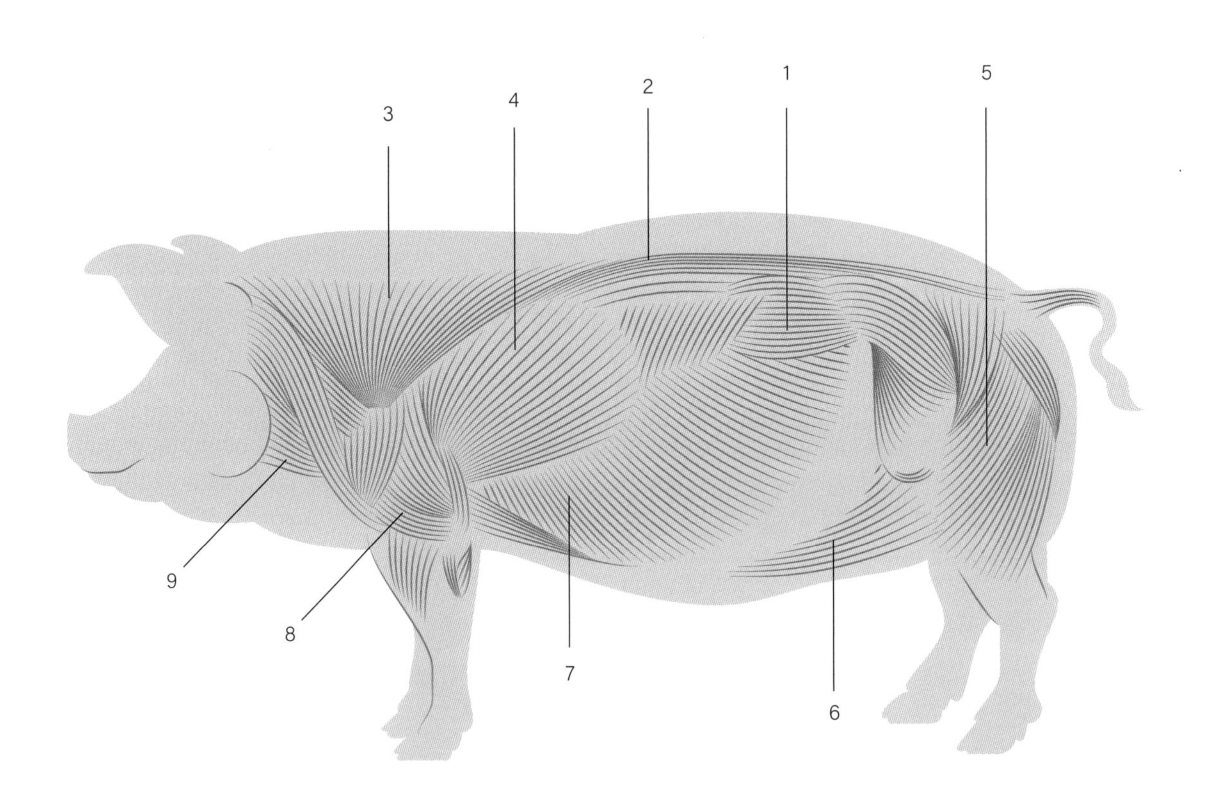

Above and opposite page, top: an overview of general North American cuts and muscle directions. Opposite page, bottom: composition of a pig.

Pork

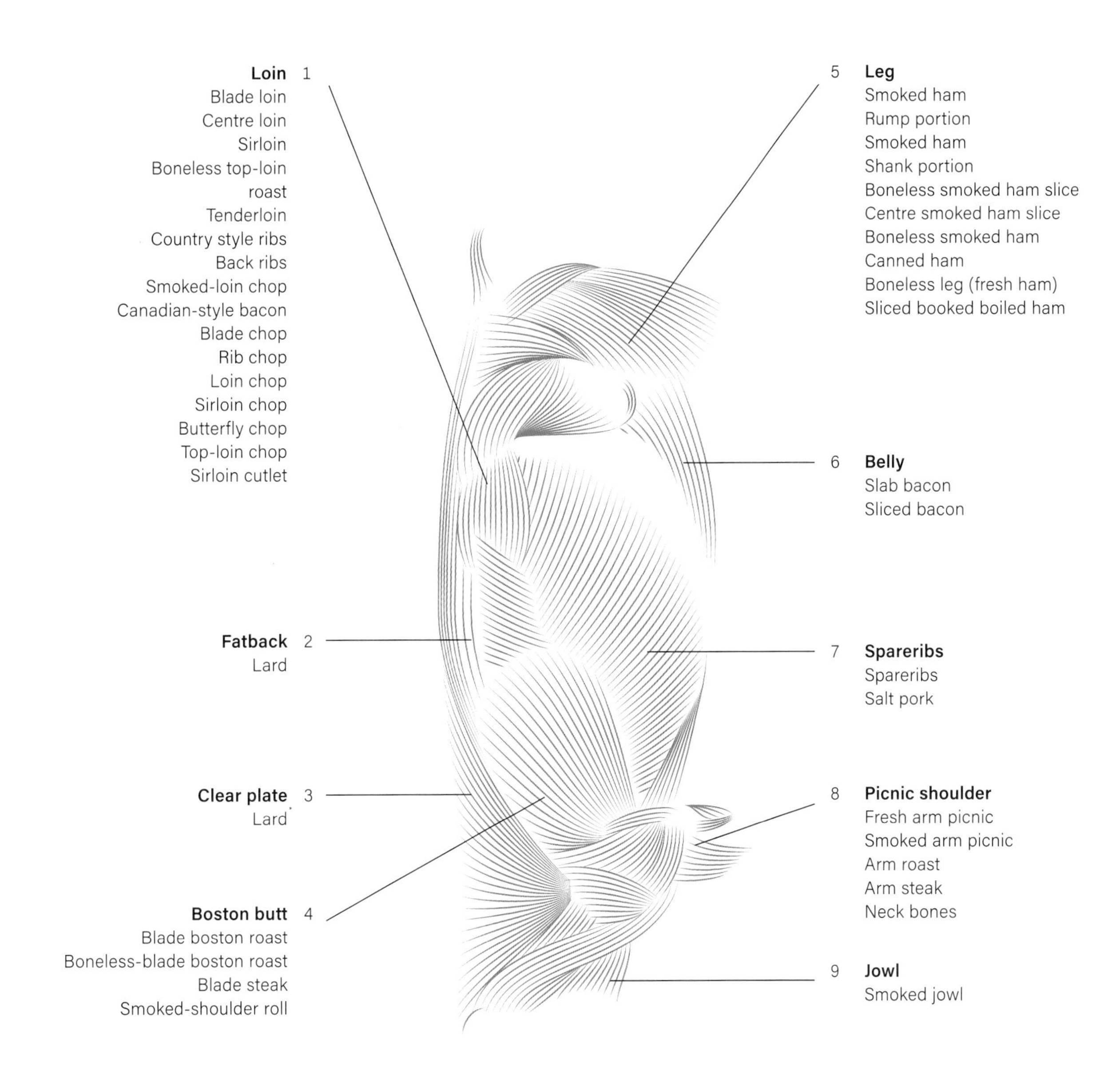

63% flesh and fat
16% bones
5% blood
8% skin
8% intestines

Average weight of a pig: 120 kg

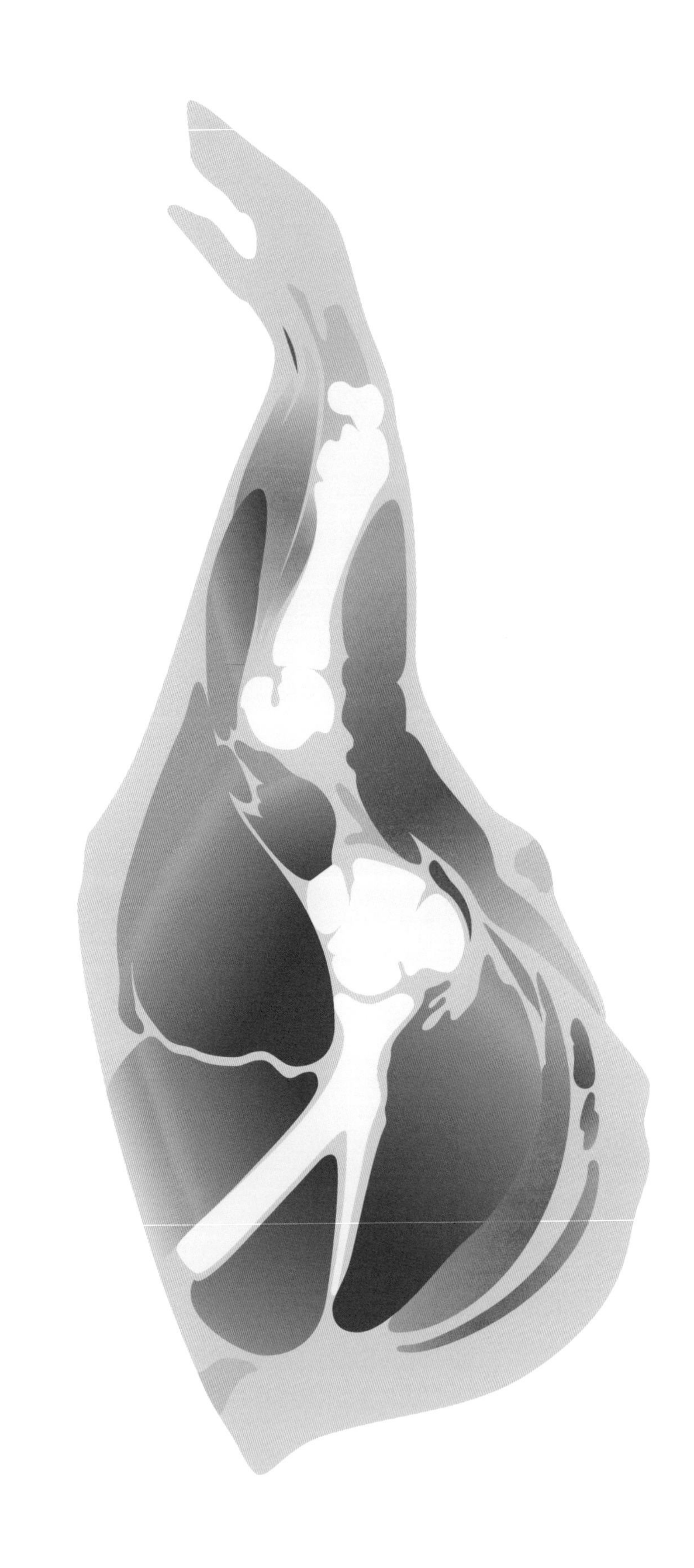

Why do chickens contain both white and dark meat, and why do the two kinds of meat taste different? Why is veal pale and delicate, whereas beef is red and robust? It all comes down to muscle fibre.

Muscle colour

The word *animal* is Indo-European in origin, and relates to the action of inhaling and exhaling: 'to breathe'. Typical of animals is movement, and muscles are responsible for that movement. Therefore, the majority of meats eaten by humans are muscle. As mentioned in the previous section, muscle fibres – the elongated cells that make up muscle – each contain an interlaced duo of specific contractile protein filaments. These compacted strands are responsible for meat's high protein content.

Muscle fibre varies in colour, flavour and type according to the task it is meant for, and there are two ways in which creatures with red and/or white muscle fibre advance. First, they move momentarily, quickly and abruptly, for example when a startled pheasant launches into the air and lands a few hundred metres away. They also move with perseverance and intent, for example when the same pheasant supports its body weight on its legs as it stands and walks. The movements described are created by one of two types of red fibres, while chicken and pheasant breasts contain white fibres. The two types differ in many biochemical respects, but the most significant difference is the energy supply each uses.

Red and white muscle fibres

White muscle fibres are designed to apply force abruptly and momentarily. They are fuelled by a small store of a carbohydrate called *glycogen,* which is already existing in the fibres and is converted quickly into energy by enzymes within the cell fluids. White muscle fibres use oxygen to burn glycogen, but if necessary they are able to generate their energy faster than the blood can deliver oxygen. As a result, lactic acid (a by-product build-up) collects while additional oxygen arrives. The restricted blood supply to the cells and the collection of *lactic acid* limit the cells' endurance, which is the reason why contracted and irregular outbursts followed by extended respite periods – allowing the replacement of glycogen and the removal of lactic acid – enable white cells to operate at their best.

Red muscle fibres are responsible for extended movements and derive their power mostly from fat. These cells get not only oxygen from the blood but also fat, which needs oxygen to metabolise. In terms of structure, these muscle fibres are relatively fine, thereby allowing them to collect oxygen and fatty acids easily from the blood. In addition, held within them are small drops of fat as well as the biochemical apparatus that can turn it into energy and that contains the pair of proteins that provide these cells with their red colour. The first of these proteins is *myoglobin,* which is related to *haemoglobin* (responsible for the colour of blood) and which temporarily stores the oxygen it derives from the blood before sending it on to the proteins that oxidize fat. Similar to myoglobin is the second type of protein, the *fat-oxidising cytochromes*, which are of a dark colour and also contain iron. The quantity in which the two proteins appear depend on the oxygen requirements of muscle fibre: as its exercise levels increase so do the protein levels. Typically, the muscles of young cattle and sheep are 0.3 per cent myoglobin by weight and therefore relatively pale, but the muscles of the constantly moving whale, which must store large amounts of oxygen during its prolonged dives, have 25 times more myoglobin in their cells, and are nearly black.

The majority of muscles have both red and white fibres (and some hybrid fibres with aspect of each), given that they are used for fast and for slow movements.

The amount of each type of fibre per muscle relates to the ways in which that muscle moves and its design on a genetic level. Some examples illustrate these differences well: migratory birds have mainly red (slow) breast muscles to allow them to fly for long durations, while domesticated birds have mainly white (fast) breast fibres, for they fly rarely. Another example is cattle, whose cheek muscles are exclusively red, allowing them continual cud-chewing, as opposed to the mostly white fibre of fast and irregularly moving animals such as rabbits and frogs.

At the Ter Weele butchery, Netherlands: collagen skins soaking, after which they will be stuffed with sausage mixture.

Skins of all Sorts

Sausage casings were traditionally various parts of the animal digestive tube. Today, most 'natural' casings are the thin connective-tissue layers of beef, pork or sheep intestines, stripped of their inner lining and outer muscular layers by heat and pressure, dried partially and packed in salt until they are ready to be filled. There are also manufactured sausage containers made from animal collagen, plant cellulose and paper. But a casing or skin can be so much more than the familiar sausages made in conventional or industrial settings. Besides such parts of the animal as the stomach or bladder, there are also entirely vegetarian skins made from beeswax, algae or leaves.

If we take the concept of a skin a little further and consider it also to be a form of packaging for the contents, there are many imaginable variations. Scientists are working on many different types of biodegradable packaging to tackle the plastic waste problem we are facing. But some new developments also have sausage skin potential: for example, the edible packaging by the USDA (United States Department of Agriculture) made from milk protein, the research for which was led by Peggy M. Tomasula, scientist in dairy and functional foods research chemical engineering. The milk protein packaging is made from the milk protein casein, and blocks up to 500 times more oxygen than currently used petroleum-based packaging. This would mean a great difference in terms of spoilage.

Additionally, in 2014, Stonyfield Organic introduced frozen yogurt 'pearls' at four Whole Foods outlets in the USA using WikiCells edible food wrappers, which were developed by the Harvard University scientist David Edwards. The skin has no flavour and can be rinsed and eaten along with the yogurt (much like a grape); it also breaks down quickly if peeled off and thrown away, eliminating the need for plastic spoons or takeaway wrappers. Yet it seems that people still want a layer of sanitary protection, since the spherical yogurts are still sold in containers. These are just two examples of contemporary packaging inventions that could easily be translated to sausage skins once they are properly developed but the aim of this book is to help create the sausages of tomorrow. Therefore in the book we are proposing casings that are available now.

The rest of this chapter explores some conventional and some less conventional but always 'ready-to-use' skins.

Intestinal sausage casings are made from the *sub-mucosa*, a layer of the intestine that consists mostly of naturally occurring collagen. These natural casings come from the intestinal tract of farmed animals; they are edible and maintain the same appearance after processing. The outer fat and the inner lining are removed during processing. Intestines are among the best sausage casings available: they are part of the same animal used for the filling, and therefore their inclusion reduces wastage and they are also the most flavourful of all the casings. Their resilient skin adapts and shapes itself naturally during the sausage-making process. We can also include in this category the more rarely used stomach, bladder and other natural animal pockets available.

Collagen casings are produced mainly from the collagen obtained from cow or pig hides, and from the bones and tendons; however, collagen can also be derived from poultry and fish. Such casings have been manufactured for more than fifty years. While collagen casings are edible, a special form of thicker collagen casing used for salami and large-calibre sausages is usually peeled off by the consumer owing to its unappealing texture. Despite its thickness, this variety of casing is permeable to smoke and moisture, as are all collagen casings. One of the advantages of collagen casings is that they give better weight and size control, and are easier to run through a machine when compared to intestinal casings. For this reason, collagen casings are more convenient than intestines when producing on a large scale.

Cellulose casings, also called peeling casings, are usually derived from cotton linters or wood pulp. Cellulose is processed to make *viscose*, a caustic-soluble xanthan salt, which is then extruded into clear, tough casings for making wieners and frankfurters – sausages that are peeled before being sold commercially.

Plastic casings made of nylon are extruded, like most other plastic products. Generally, smoke or water do not pass through the casing, so plastic is used for non-smoked products where high yields are expected. The inner surface can be laminated or co-extruded with a polymer with an affinity for meat protein that causes the meat to stick to the film, resulting in some loss when the casing is peeled but higher overall yield due to a better control of the moisture content.

Fabric casings are traditional, but have been in use only occasionally, mainly when the desired sausage shape did not correspond to any of the animal's array of natural pockets. Nowadays, fabric is used for special occasions, decorated with elaborate prints such as pine cones or Father Christmas. And while this type of skin is inedible, there are casings available that contain a coating on the underside, for example ones with a layer of crushed pepper that gives the sausage a special flavour.

Fibre reinforced casings are made of cellulose and are reinforced with a faser fibre that strengthens the casing to make it suitable for bigger sausages. The casing is non-edible, but often has an 'easy peel' layer to allow for simple removal of skin from sausage.

Wax is the natural material in which bees store their honey and in which they breed. The primary reason for waxing is to prevent water loss. As such, this technique is often used for citrus fruits, apples, and other fruit to prevent shrinkage and spoilage, and to improve appearance. Although waxing is not a common technique for sausages, there are traditional *leberwurst* (German for 'liver sausage') that are dipped in fat to create an airtight coating, so as to prevent the discolouring of the liver. The interesting feature of the waxing technique is that it seals the content completely from oxygen, thereby preserving it for an extended period, yet it remains entirely natural.

Algae or seaweed has been rising in popularity worldwide recently. The production and processing of nori seaweed is an advanced form of agriculture. Its biology, although complicated, is well understood, and this knowledge is used to control the production process. The processing of the raw product is mostly carried out by highly automated machines that recreate traditional manual processing steps accurately, but with a much-improved efficiency and consistency. The final product is currently used mostly for the wrapping of *sushi*, but it has good potential to be seen in future as an edible, breathable skin for different purposes too.

Leaves – both edible (lettuce, cabbage, vine) and inedible (banana, fig, bamboo) – are thin and broad. They are used as wrappers to contain, protect and aromatise fillings of meat, fish, grains, and other foods. The leaves are often blanched first, to make them flaccid and pliable.

Gelatine casings can be made of animal-based gelatine or more plant-based ingredients, such as algae and fruit. They can make a non-breathable skin that seals the sausage from oxygen but is still edible. Yet while gelatine is widely used in terrines, to cover tarts and in desserts, its use as a sausage casing remains little known (but not unthinkable!).

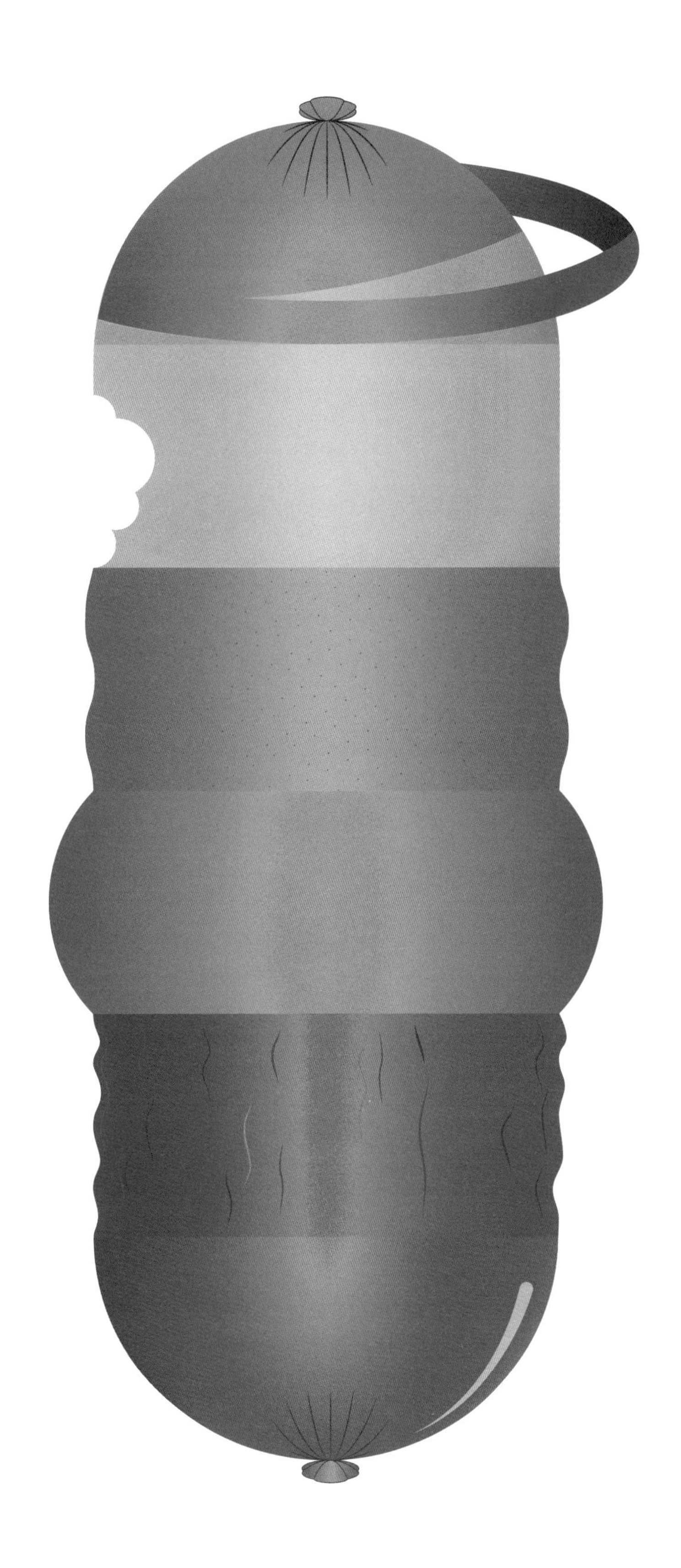

Different types of skin, from top: inedible peel-off, edible, natural, artificial, breathable and non-breathable.

Method

Preservation & Glue

While forming only a small percentage of the sausage mass, these ingredients – or aspects, if you will – are the ones that transform bits and pieces of meat and other elements into the sausages that we know and love. The following essay is a more in-depth look into these components.

45 Production Techniques
51 Types of Sausage

Preservation

Preserving food was once a question of life or death: it was necessary in order to survive winter and other precarious times. Thanks to refrigeration and global shipping, preservation is no longer a matter of survival, but the techniques and recipes for doing so remain cherished and of cultural importance. Besides, our different ways of preserving food provide unique flavours and a diversity of textures that we have come to love dearly: cheeses, hams, pickles and, of course, sausages. From an historical point of view, the most common approach to preservation is to deprive the microbes of the water they need to thrive, either by dehydrating the food or by adding salt or sugar, which take water away from microbes by chemical means. Another way to hinder bacterial growth is to lower the pH of the water contained in the food by adding vinegar or some other acid, or else by encouraging the growth of certain desirable bacteria on food that can out-compete the microbes that cause spoilage. Most other methods of preserving, such as smoking and freeze-drying, work by removing water or adding compounds that directly suppress bacterial growth, or by some combination of both. Canning involves sterilizing food so that all enzymatic activity and bacterial growth is halted. Freezing, on the other hand, does not eliminate enzymes or bacteria, but instead it effectively halts their activity to prevent spoilage. Heat destroys all bad microbes, but if used as a preservation method, it will always need an additional method to keep the spoilage microbes from restarting their process.

Removing moisture. Microbes need water to survive and grow, so one traditional preservation technique has been the drying of meat, originally by exposing it to the wind and sun. Nowadays, meat is dried by salting it briefly to inhibit surface microbes and then heating it in low-temperature convection ovens to remove at least two-thirds of its weight and 75 per cent of its moisture (the presence of more than 10 per cent moisture may allow *penicillium* and *aspergillus* moulds to grow). Because its flavour has been concentrated and its texture is unusual, dried meat remains popular. Modern examples include American *jerky*, Latin American *carne seca*, Norwegian *fenalår* and southern African *biltong*; their texture can range from chewy to brittle. Two refined versions are Italian *bresaola* and Swiss *bündnerfleisch*, which are made from beef. They are salted and sometimes flavoured with wine and herbs before a slow, cool drying period of up to several months. Both are served in paper-thin slices.

Salt. The prevailing substances added to food in order to preserve it are sugar and salt, which stop microbes and cells from causing additional decay by drawing out water from them and binding it to themselves.

When **acid** is used as a preservative, food is conserved in an edible antimicrobial liquid with a pH lower than 4.6, which is sufficient to kill most bacteria. This is a common technique in pickling vegetables, but is also used for sausages and eggs. In the USA, for example, there is a pickled sausage that does not require refrigeration and is usually made with a smoked or boiled sausage plunged in a boiling brine – made of vinegar, salt, spices and often a pink colouring – and put in pickling jars. The result of this technique is, as might be expected, a sour-flavoured sausage with prominent notes of the spices used, which in turn cause it to lose much of the more delicate original sausage flavour.

Good bacteria. Preservation does not solely mean keeping a food forever; it also means generating a delicious flavour that did not exist in the original item. This process is where fermentation can really shine.

> *Fermentation is the intentional cultivation of helpful microbes in order to prevent the growth of spoilage microbes.*

There are many ways to start a controlled fermentation. One of the simplest ways is to take a small batch of previously preserved food and add its microbes to the fresh food that needs preserving. Those microbes used for preserving the food are simultaneously busy creating all sorts of aromas that did not exist before. Hence, the acidity that they develop gives a wonderful freshness to pickled and preserved food. Another way to start a successful fermentation is to give the good microbes an advantage over the spoiling microbes by changing the chemical conditions on the surface of the food. For example, sprinkling salt onto the food will draw out moisture that will feed the desired microbes, so that they receive an advantage over the spoilage microbes. We know that milk can be transformed into long-keeping and flavourful cheese by removing some of its moisture, salting it, and encouraging harmless microbes to grow in and acidify it; similarly, meat can be treated to the same effect.

Fermented sausages are thought to have their origins in prehistory, from the practice of salting and drying scraps of meat to preserve them. Inside the wettish mass salted scraps are squeezed together and microbe-filled surfaces are created. Further, under these conditions, bacteria that are tolerant to salt and are able to grow in the absence of oxygen will multiply. These bacteria tend to be *leuconostocs* and *lactobacilli* (as well as the similar *pediococci* and *carnobacteria*), which also thrive in cheese that is low in air and high in salt content. And they also create acetic and lactic acids, which decrease the meat's pH levels from 6 to 4.5–5, thus it becomes less of an inviting environment to decay-causing microbes. As the drying process continues, the levels of acidity and salt increase thus making the sausage less susceptible to decay.

Smoke from the burning of wood and plants has been, owing to its chemical complexity, an aid to the preservation of foodstuffs for as long as fire has been in use by humans. Its hundreds of compounds have different effects: a pleasing flavour, the slowing down of rancidity or the oxidation of fat, or the stopping or slowing down of microbe development. Yet it is often combined with drying or salting, given that its impact is merely on a food's surface, and works especially well with meat, which tends to become rancid.

Freeze-drying was first developed to make a type of jerky called *charqui* by natives of the high altitude and dry environment of the Andes, the moisture being removed naturally from meat on hot days and diverted into ice crystals on cold nights. The honeycombed tissue that remained uncooked was created, but it easily reabsorbs moisture during cooking. Modern techniques see a quick freezing and vacuuming of the meat, followed by gentle heating so the water is sublimated. As a type of drying, it does not cook and compact the meat, therefore it allows for thick cuts to be used.

Heat is one of the most important preservation techniques because it aims to destroy, or partially or totally inhibit, bad microbes and enzymes. This technique has a major effect on the taste, texture and nutritional values of the food. There are different stages of heat treatment in food processing for preservation. Sterilization is the heating (most often above 100 °C) for a specific period of a certain food in order to destroy bad microbes and enzymes. The amount of heat and the time required depend on the type of food and the microbes/enzymes it contains.
A similar method is pasteurization, where the heat can be much lower, around 70–80 °C. This alters the flavour and texture of the food much less, but also lowers its potential shelf life.
A common heating technique in the making of sausages is poaching: a water bath that stays at a constant temperature of around 80 °C, enough to cook the meat but not extract too much moisture from it. Mortadella and blood sausage, for example, are treated in this way.

No oxygen. Simply removing contact with oxygen is not enough to preserve a food item for very long. However, when combined with heat, its shelf life can be increased almost indefinitely. This way of preserving food removes all its microbes. The foodstuff first needs to be sealed in a container, to protect it from the microbes that surround us. Then the container is heated up to a certain degree, thus making sure all the spoilage microbes have been destroyed. Some similar processes do not involve this extended heating technique, but instead make use of combinations with other preserving ingredients. This is the preservation process for most cooked sealed sausages, again such as mortadella.

Cold storage such as the refrigeration of food is a gentle method of food preservation: it does not destroy bad microbes or inactivate enzymes, but rather slows their deteriorative effect. It has minimum adverse effects on the taste, texture and nutritional value of foods. Refrigeration has a limited contribution towards preserving food, for it extends shelf life by a few days only. Most spoilage microbes prefer warmer temperatures, but there is a group of microbes called *psychrophiles* which will grow at refrigerated temperatures thus causing spoilage. Freezing, on the other hand, has a much longer contribution towards preserving food, but it also affects the taste, texture and nutritional values of food.
By freezing food, the growth of spoilage microbes is almost entirely stopped, but enzymes will maintain a certain level of activity, thus causing it to spoil.

Glue

A key aspect of sausage making is creating a cohesive gel that will hold the sausage together. Indeed, the casing exists mostly to hold the sausage together until the gel sets. This glue, or gel, is also referred to as the binding of the sausage. How one binds a sausage depends on what type of ingredients they are using and how they handle them. For example, meat contains a natural binder called myosin that sets into a gel when it is heated. However, not all meats have great binding powers: most organ meats, for example, have poor binding of their own, so they would benefit from additional glue. Blood and egg, on the other hand, are excellent binders because they coagulate when heated up, thus setting into a solid form and binding everything around them together.

An interaction between molecules that produces systems in which liquid is held and mass is created makes for a successful binding. Gels respond to the heat–cold cycle in one of two ways: either by thermo-reversibility (when a gel becomes liquid again on being heated) or thermo-irreversibility (when a gel does not return to a fluid state on being heated). A good example of thermo-irreversibility is egg, for once its proteins connect to form a network, they remain so; thermo-reversibility is exemplified by gelatine, which network can disconnect upon heating. The majority of binders need water to be diluted with them to provide easy motion, but otherwise they often have different chemical requirements. So, most binders require a *coagulant* (gel inducer) such as calcium to create a lasting bond, which provides the acid in certain cheeses, for example.

What heat does to an egg is to cause the increasingly rapid movement of its molecules, causing them to collide repeatedly and more powerfully, and finally the links between the chains of protein disconnect, only for the proteins to open up and connect in a new, three-dimensional system. While the water ratio is much higher than the protein, it is stopped from flowing easily owing to the different small areas of the protein system into which it is now divided, and the egg becomes solid yet moist. Moreover, the egg albumen, at first sheer, is now opaque, owing to light bouncing back from the thick grouping of large protein molecules. The principle of bridging the distance between the proteins and encouraging a connection between them is relevant in different processes that make eggs more solid, be it creating a foam (through beating) or pickling (through pickling in salt or acid).

The Victorian satirist Samuel Butler awarded the egg priority when claiming that a chicken is just an egg's way of making another egg.

Generally, binders can be divided into the following categories: animal protein binder, hydrocolloid binder, starch binder and fibre binder.

Proteins can be found in all sorts of animal tissue, but the following are important as binding proteins. Myosin, as mentioned before, is a protein that leaks out of meat when it is minced and packed together. When heated up, the myosin turns into a gel, thus binding the sausage together. The amount of binding generated depends on the animal and the cuts of meat used. The best way to let myosin do its work is to cut the meat into small pieces, or mince and then apply salt to it, to maximize the meat surface from which the myosin is drawn.

Other glues that perform well by being in a protein-rich environment are different forms of enzymes such as transglutaminase, rennet, and other protease enzymes. These can glue protein-rich liquids or substances together by cross-linking or bonding amino acids that, without the enzyme, would not have any binding power.

Gelatine is a protein derived from animal-based collagen (from hides, bones and cartilages) and is a translucent, colourless, brittle (when dry) and flavourless gelling agent. It is composed of elongated protein molecules. These molecular strands have an affinity for their own kind, so as they cool, they nest together to form a three-dimensional mesh, trapping water molecules in the process. Although it is a protein, gelatine can be used in the same way as a *hydrocolloid* (a starch or plant gum used to thicken or gel water-based liquids): by soaking it in water before adding it to a liquid and heating it up, or by adding it to a hot liquid directly.

Starch is the most familiar type of thickener, and contains two types of long tangled molecules (called *polymers*): *amylose* and *amylopectin*. Both are large chains of *glucose*, a type of sugar. However, they are not detectible by the sweetness receptors in our tongue and therefore are not sweet in taste. Amylose is better for forming gels, whereas amylopectin binds up water and thickens better. A starch always needs to be hydrated with water, but can be dissolved in oil or fat before being mixed with water. This functions like the flour in a roux, where the hydration happens when the milk is added and thickens properly when heated.

Like starches, **hydrocolloids** are always hydrated with a water-based liquid (hydro = water). When the hydrocolloid is dispersed in water, colloid particles are spread throughout and depending on the quantity of water can create different states; gel or liquid. Hydrocolloids can be either irreversible (single-state) or reversible. For example, *agar,* a reversible hydrocolloid of seaweed extract, can exist in both gel and liquid states, and can alternate between states with the addition or elimination of heat.

Fibre is defined as the material in our plant foods that our digestive enzymes cannot break down into absorbable nutrients. Thickening a liquid with fibres essentially means adding microscopic chunks of plant or animal tissue to a liquid so they are suspended, thus giving an impression of thickness by interfering with the flow of the liquid.

Emulsion. There is another way of disturbing this flow, similar to the addition of fibres, by forcing two incompatible liquids together. Water molecules and oil molecules do not mix evenly with each other because of their very different structures and properties. However, if we use a whisk or blender to force a small portion of oil to mix into a larger portion of water, the two form a milky, thick fluid. Both the milkiness and the thickness are caused by small droplets of oil, which block light rays and the free movement of water molecules. Such a mixture of two incompatible liquids, with droplets of one liquid (dispersed phase) dispersed into the other liquid (continuous phase), is called an emulsion.

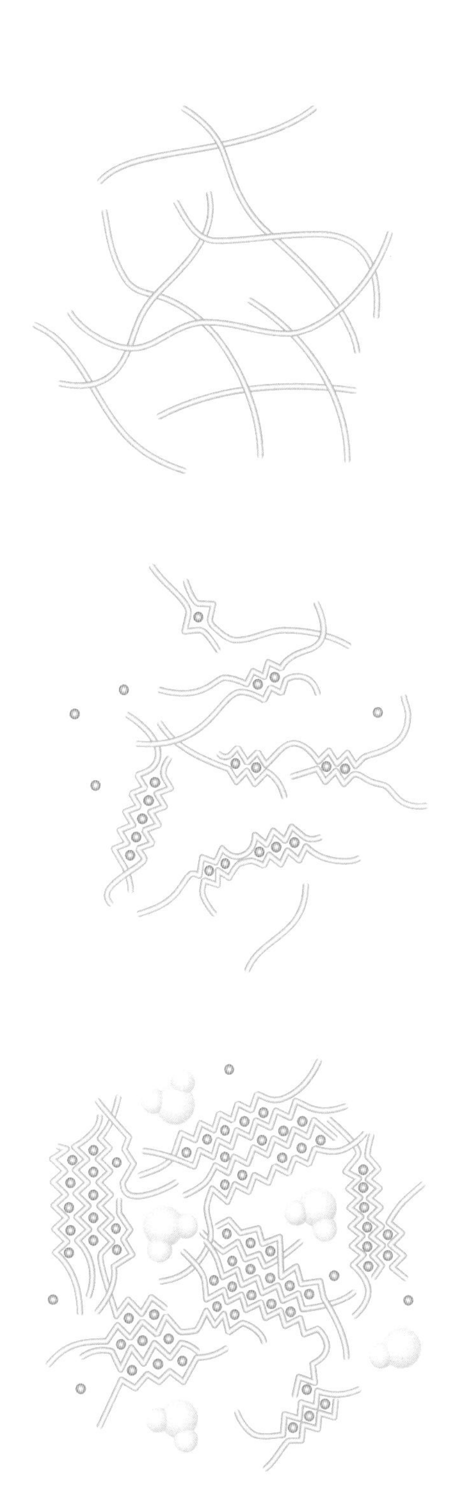

Glue requires a neutral companion to build its connections around. In this case, strands of protein wrap around spheres of calcium, thus creating chambers in which water can be captured.

At the Ter Weele butchery, Netherlands: an adventurous mixture of blanched mealworms with pork fat and a bit of cow's milk – a trial for insect pâté. One of many tests made at the butchery during the making of this book.

Production Techniques

The handling of the ingredients and the making of the sausages can be categorized into a few general procedures: mincing, extruding, filling and linking. These techniques are done from very small to very large scales and by different-looking machines, but the technique is often the same, as shown in the following selection of schematic sausage-making machines.

Mincing is the process of chopping different cuts of meat down to the appropriate size according to the desired result, such as a sausage, burger or meat loaf. To mince meat properly there are a few guidelines to follow. For example, always cool the substance to around 0 °C so it will mince well and thus prevent the fat from melting and smearing. Moreover, one should mince the different ingredients separately from each other, since they often differ in structure and therefore need different forces to be minced evenly.

Extruding. In the extrusion process, raw materials are first minced to the correct particle size: coarse, medium, fine or smooth. The mixture is then combined with other ingredients that are added depending on the target product. These may be seasoning, nuts or seeds for texture, or fat or liquid for juiciness, and more. When the mixture – the *extrudate* – is ready, it can be passed through the extruder, which consists of a large, rotating screw that fits tightly within a stationary barrel, at the end of which is the extrusion mouth. The extruder's rotating screw forces the extrudate towards the mouth, through which it then passes into the casing.

Filling a sausage is necessary when the stuffing is too liquid to be extruded. Instead of stuffing the casing horizontally, filling happens vertically, under the force of gravity and of pouring. The casing is held open and the liquid mixture is poured in from the top. When the desired sausage size is reached, the casing is closed and made airtight. Often, it is then put into a water bath of around 80 °C to coagulate the mixture and make the interior set. This is important to ensure the sausage can be handled and cut into for consumption without the contents running out of the casing.

Linking refers to the portioning of sausages. The linking of sausages is a precise procedure done by hand or industrially by machine. This process of making individual portions is important for the further handling of a sausage. A long string of sausages is more likely to break if the contents are divided unevenly.

There are many ways to link sausages, and the way this is done will often define the characteristics of the sausage variety.

When done by hand, the easiest method is to pinch and twist. More elaborate techniques involve folding links one over another to shorten the length of the chain, making it easier to hang for setting, drying and smoking.

Mincing

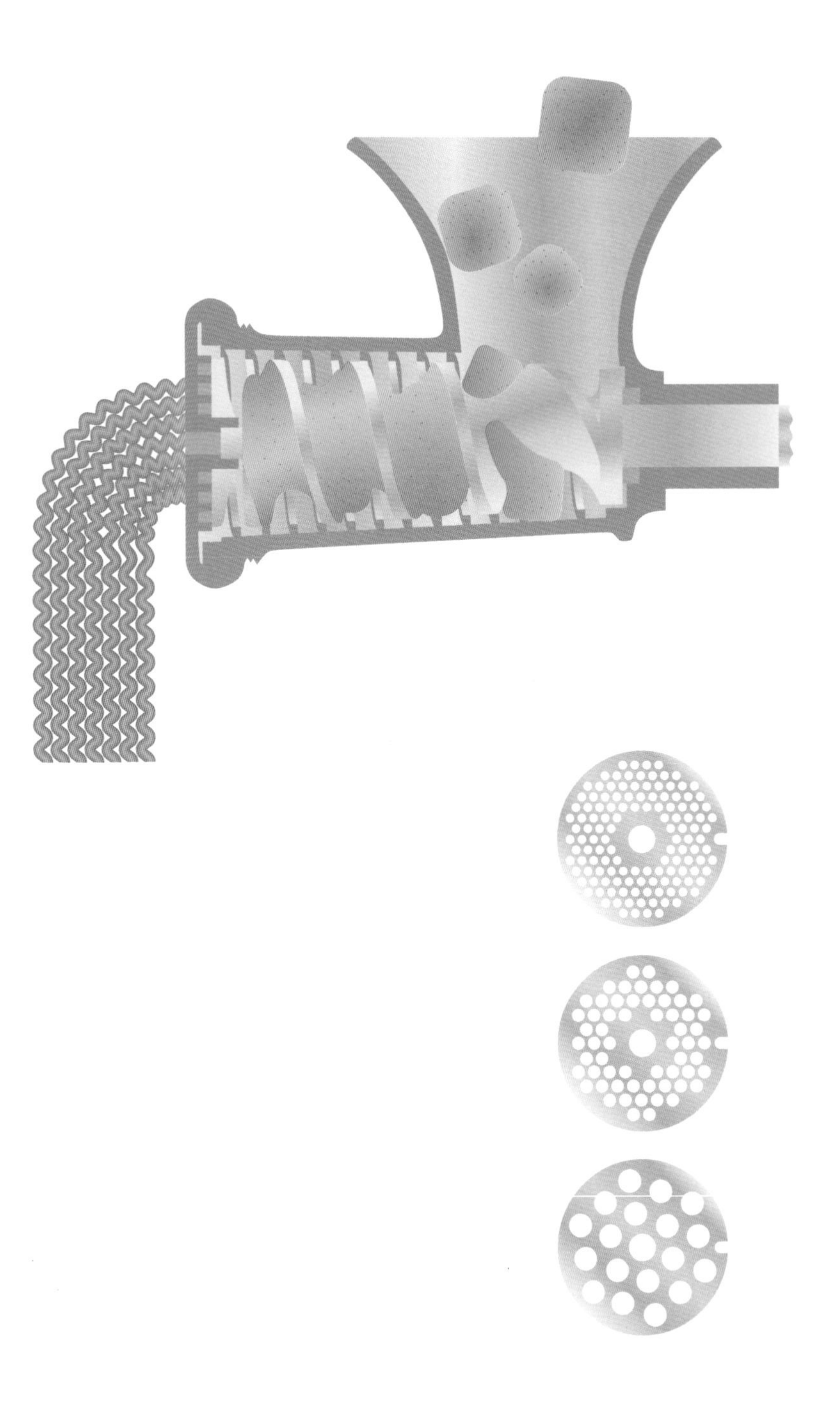

This and following spread: illustrations of machines whose production capacity varies from artisanal to semi-industrial sausage production. While the machines are of different scales, the technique remain the same.

Amateur mincer:

1 There are three mincing plates to choose from, according to the desired meat coarseness.
2 The meat and other ingredients are put into the top of the machine, one ingredient at a time.
3 The rotating screw pushes the meat towards the cutting blades and the mincing plates.

Filling and Extruding

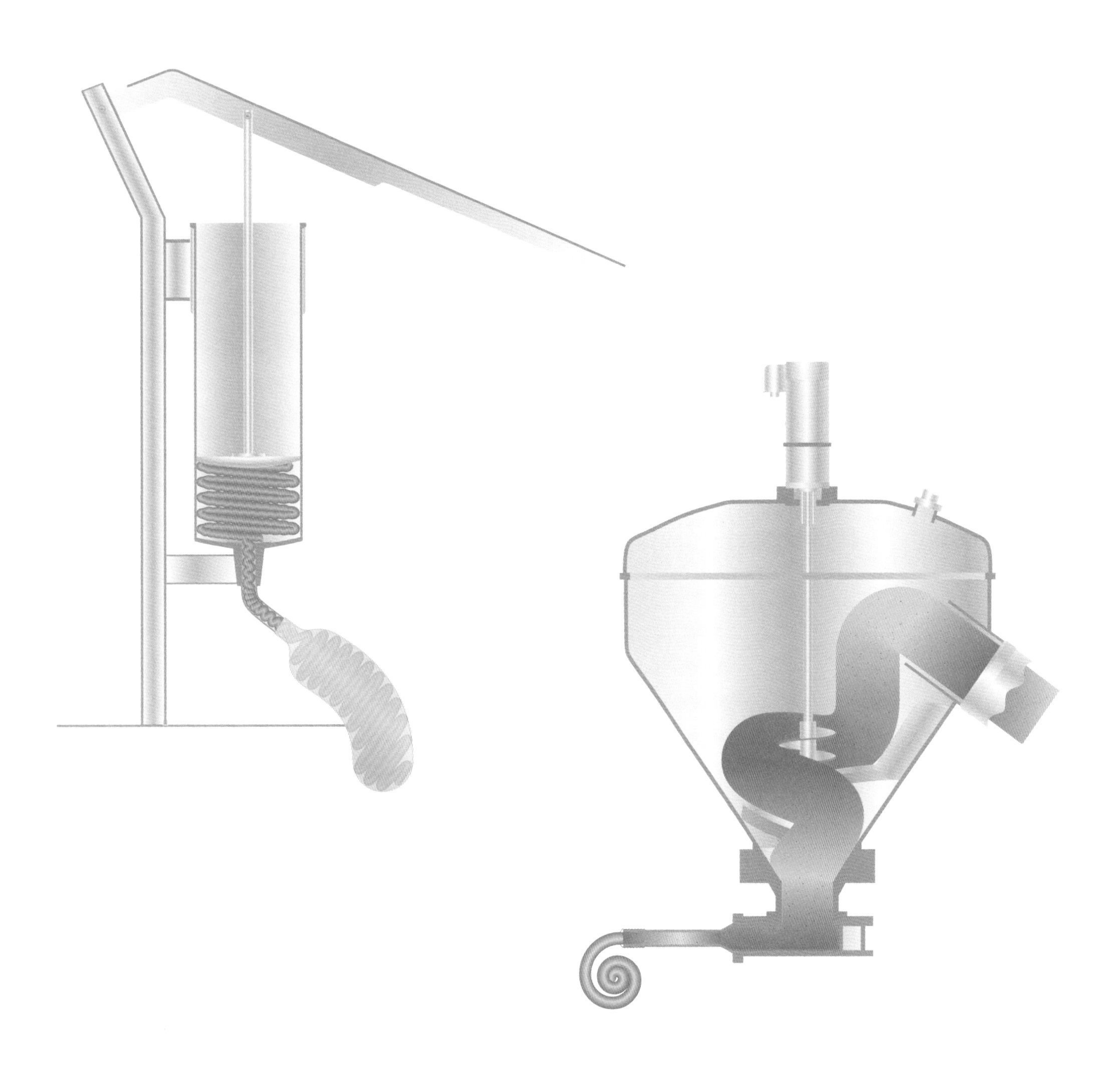

Filler (used by artisanal butchers):

1 The meat paste or liquid mass enters the machine from the top.
2 The casing is attached or held on the exit mouth.
3 The casing is filled by gravity, and if necessary the mass is pushed down by the handle.

Semi-industrial meat extruder:

1 The meat mixture – the extrudate – enters the machine and is pushed down by the screw to remove most air pockets.
2 The casing is rolled up on the extrusion mouth.
3 The extrudate exits the extrusion mouth into the casing; this is often guided by hand.

Linking

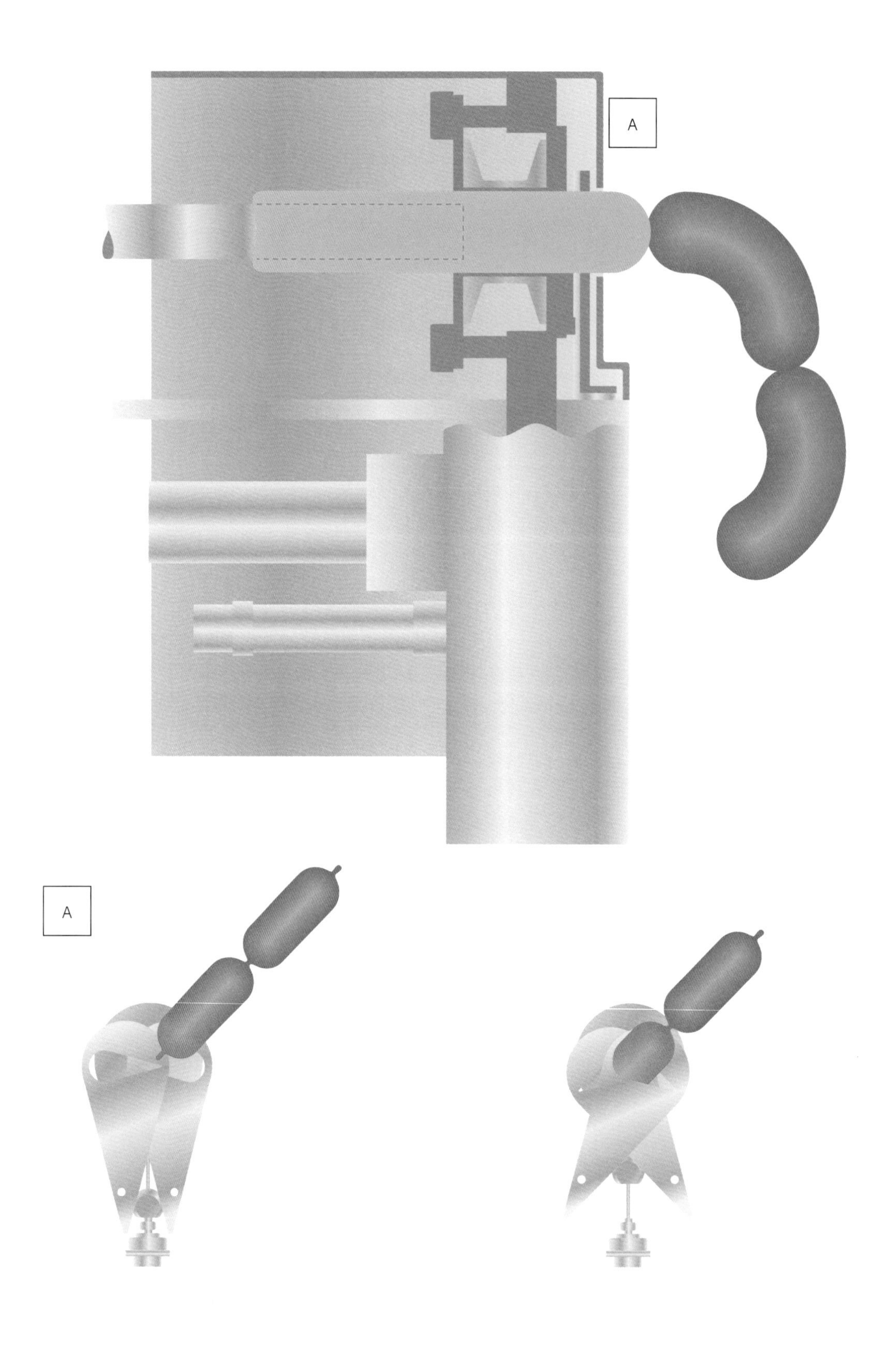

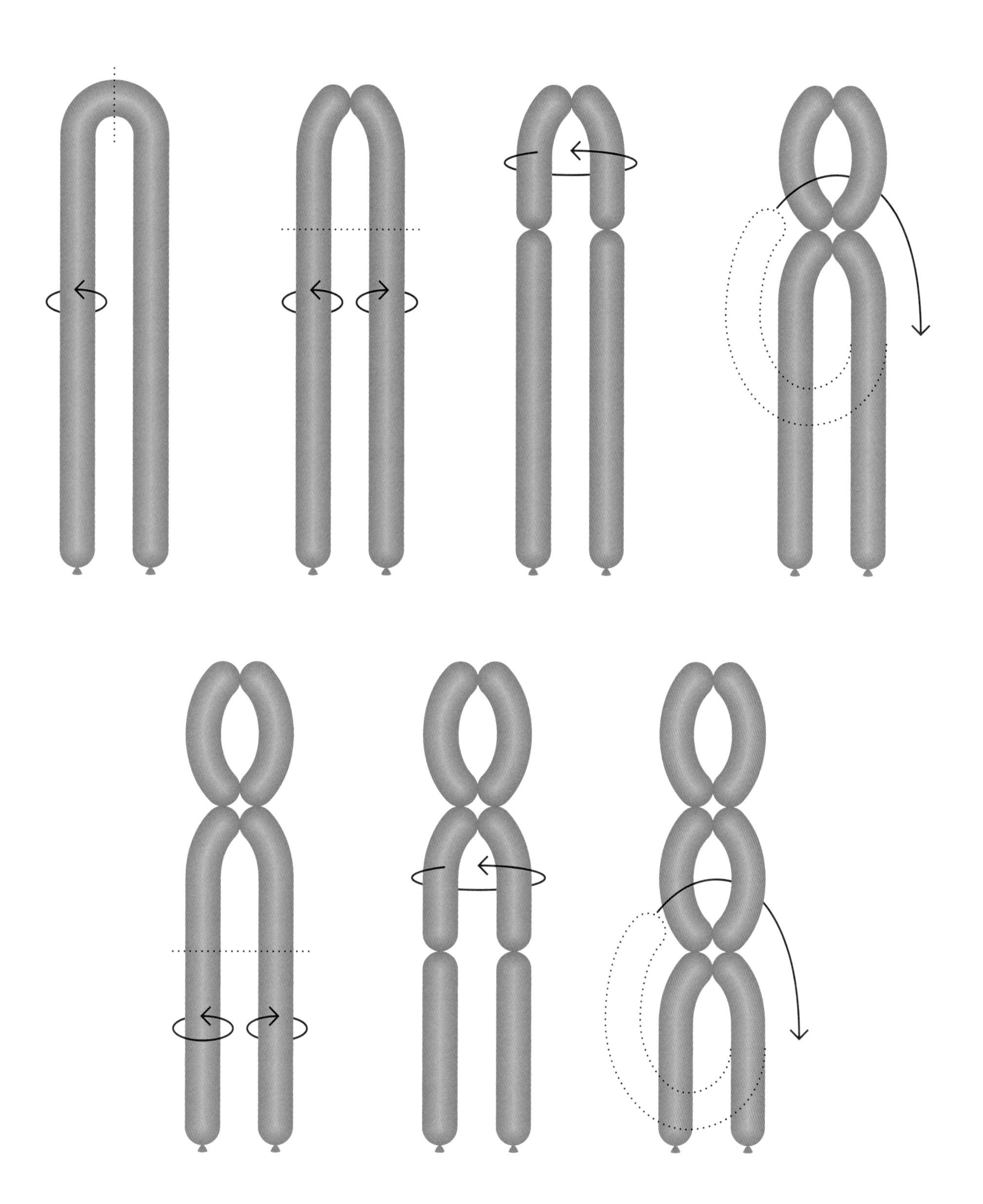

Semi-professional sausage linker:

1. The casing is rolled up on the extrusion mouth through the linking plates (A).
2. The extrudate exits the extrusion mouth into the casing.
3. The linking plates, just after the extrusion mouth, open and close into a ring that presses the meat-filled casing into a pre-set sausage length.

Common technique for manual linking:

1. Twist the middle of the sausage string to make the first separation.
2. Choose the desired sausage length and twist into individual links.
3. Twist both the links together.
4. Take one of the sausage legs and pass it through the sausage circle.
5. Repeat with the rest of the sausage until it is fully linked.

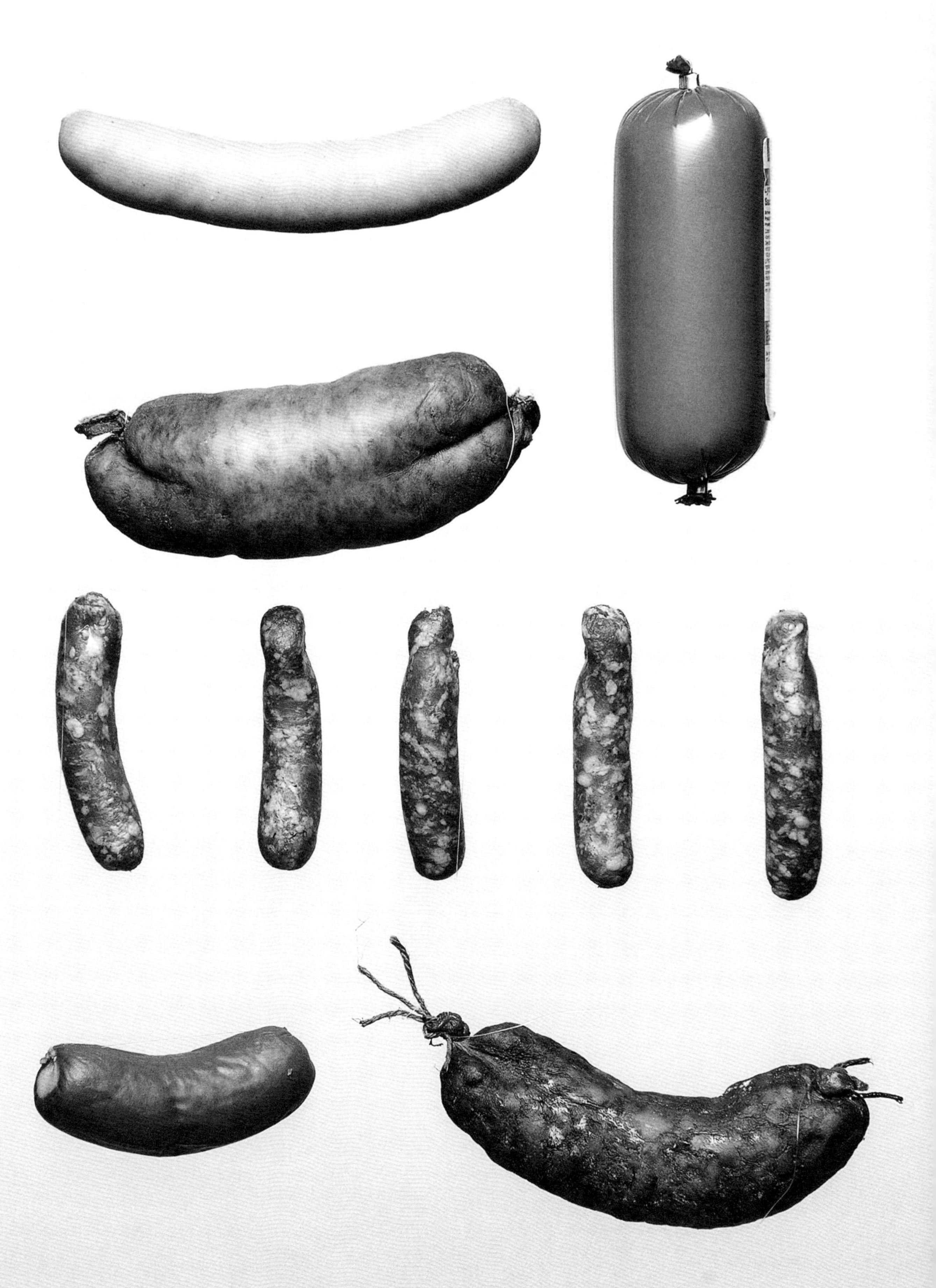

A selection of sausages from a supermarket and a butchery in Lausanne, Switzerland.

Types of Sausage

As discussed in the anatomy of the sausage on page 9, the definition of what a sausage is can be viewed very broadly. For example, the French *andouille* or *andouillete* is a meat preparation whereby the intestines are filled with layers of intestines, stomach, and other offal, giving the sausage a unique and layered pattern. Or take the Italian *zampone*, a sausage that uses a very different type of casing from most other sausages; this is a hollowed-out pork foot filled with sausage meat and traditionally eaten with lentils. On the Balearic Islands *sobrassada* is a common sausage that consists of a heavily spiced (often with paprika), raw, cured meat paste extruded into a casing; it is eaten by spreading the paste on bread or toast. Sobrassada is part of the sausage meat products prepared in the laborious but festive rites that still mark the autumn and winter pig slaughter known as a *matanza*. The chemical principle that creates sobrassada is the dehydration of meat under certain weather conditions (high humidity and mild cold), which are typical of the late Balearic autumn. Another peculiar way to interpret a sausage is the Spanish *lomo embuchado*, which is essentially an entire pork loin, rolled through pimentón and put in a casing. Lomo embuchado, also known as *lomo curado*, is then air-dried for around three months before this delicacy is eaten, often sliced very finely.

These are exceptions; most sausages fall into just a few categories: fresh, cooked, dried, blood, and cured. But with so many different types of sausage out there, not all will fit into these categories or descriptions. Therefore, each drawing in this section is inspired by one type of sausage from one category.

Fermented sausages are among the most flavourful kinds thanks to the bacteria that break down bland proteins and fats into smaller, intensely savoury and aromatic molecules. Fermented sausages develop a dense, chewy texture that has a distinct, slightly tangy, mouth- watering taste. When curing is done correctly, the enzymes and bacteria are beneficial for, and easy to digest by, humans. There are many kinds of food products whose shelf life is increased greatly by fermentation, for example when milk becomes cheese. When sausages are mainly fermented and not additionally dried, their shelf life is not extended by much but their flavour profile (a method of judging the flavour of foods) score is 'delicious'.

Fresh sausages are made of minced or chopped meat mixed with salt, spices, and sometimes other ingredients such as breadcrumbs, or (for gluten-free varieties) oat flakes or rice. The mixture is often put into an intestinal or collagen casing and always cooked before eating. The British often refer to fresh sausages as *bangers*, in reference to the popular dish bangers and mash (sausages and mashed potatoes).

The term *banger* dates as far back as 1919, and is believed to have been popularised during World War II, when scarcity of meat led many sausage makers to add water to the mixture, making the sausages more likely to explode during cooking. They are a famous essential component of a full English or Irish breakfast.

> *It is believed that there are over 470 different types of sausage in the UK.*

Some are made to traditional regional recipes – such as those from Cumberland or Lincolnshire – and increasingly to modern recipes which include fruit such as apples or apricots in the mixture; others are influenced by European styles such as the *Toulouse sausage* or *chorizo*.

Dried sausages are perhaps one of the oldest sausage varieties. Throughout history, it has been a major challenge for mankind to preserve meat against biological spoilage. The earliest methods, which date back at least 5,000 years, use physical and chemical treatments known to make meat inhospitable to microbes. For example, drying meat in the sun and wind or by the fire removes enough water to stop bacterial growth. Salting is also one of the oldest preservation methods available, and since salt helps to extract moisture from meat, this method was often used in conjunction with smoking. Heavy salting with partly evaporated seawater, or rock salt, or the ashes of salt-concentrating plants also draws vital moisture from the cells. Often, dried sausages are partly cured. Moderate salting allows the growth of a few hardy and harmless microbes that help exclude harmful ones. Some of our most complex and interesting foods have developed out of these crude methods. Nowadays, dry-cured hams and fermented sausages are a treat, but until relatively recently they were essential for travelling since the sausage did not need refrigeration and could be taken on long journeys and thus provide necessary protein and energy.

Cooked sausages are heated through, with or without smoke, before being sold. This contributes to the coagulation process of the mixture, and also means they can often be eaten cold. Cooked, emulsified sausages are mixed thoroughly so that the meat and the fat form a smooth, emulsified batter that sets into a uniform gel. These sausages are perfect for using scraps, since the texture is so fine that the smallest parts of meat become one with each other. Emulsified sausages are often named after their origin, for example *frankfurters* from Frankfurt, *wieners* from Vienna and *mortadella bologna* from Bologna. The latter, from Italy, is made of a very smooth pink pork mixture with cubes of soft pork fat, and is studded with black peppercorns (sometimes also with pistachio nuts, mostly outside of Bologna). It is enjoyed thinly sliced, then folded and layered thickly in a sandwich. Mortadella is often very big, because it stays fresh for a long time owing to its carefully pre-cooked production.

> *There is a record of a sausage made by mortadella brand Veroni in 1989 that weighs over 1,300 kilogrammes and is 5.8 metres long.*

Blood sausages are found worldwide. Pig's, cow's, sheep's, duck's or/and goat's blood is used, depending on different countries. In Europe and North America, typical fillers include meat, fat, suet, bread, cornmeal, onion, chestnut, barley, rye and oatmeal. In Spain, Portugal and Asia, potato is often replaced with rice. In Kenya, fillers are fresh minced goat or beef, fat and red onion.

The Finnish *mustamakkara* is a blood sausage made with pork, pig's blood and grains of rye. This popular sausage is often bought at a market and eaten as a late breakfast or lunch, and traditionally enjoyed with lingonberry jam and milk.

Fermented sausage

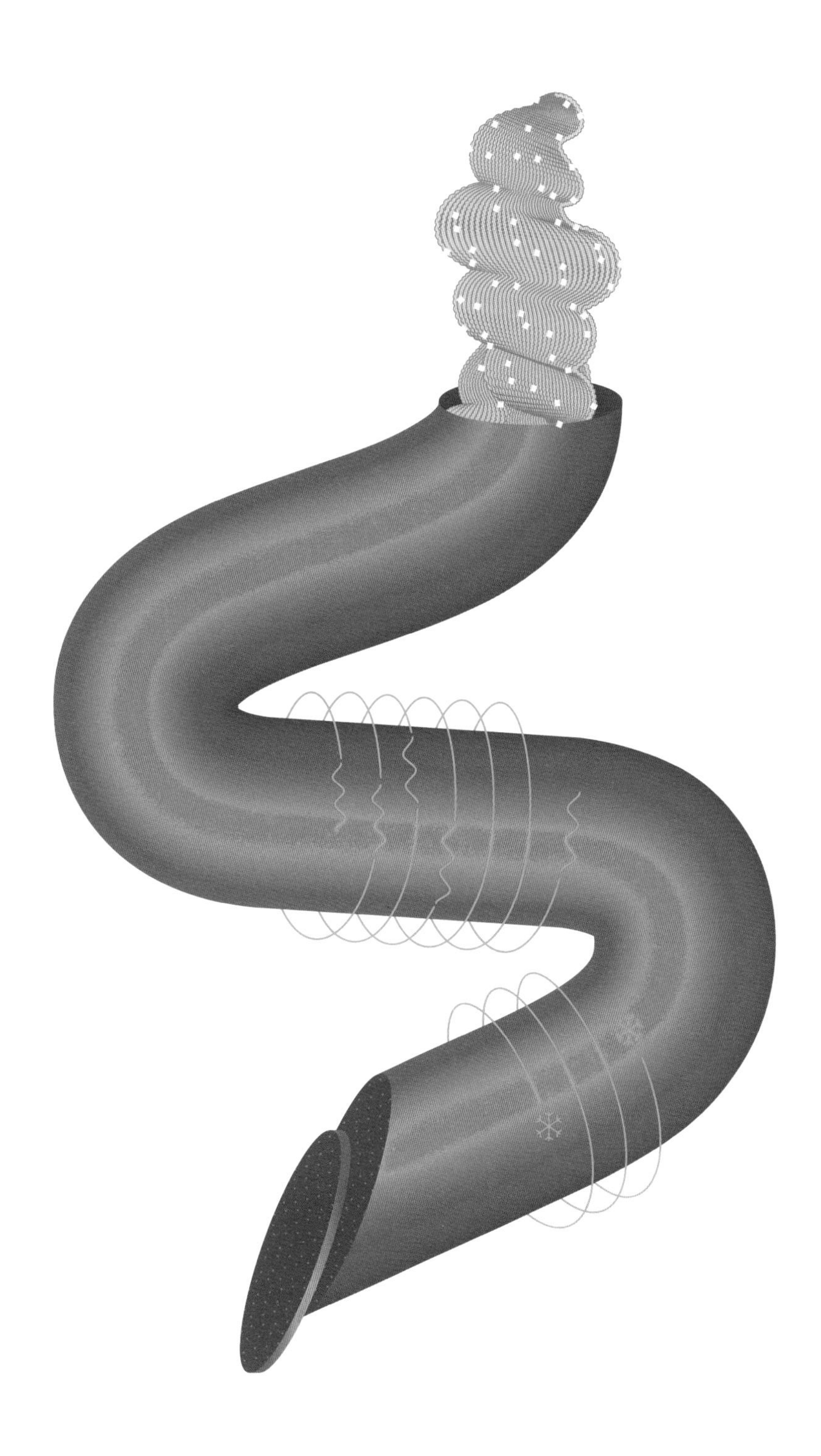

A traditional Dutch variety of fermented sausage, the *osseworst* (literally 'ox sausage') was originally made from ox meat, but now generally from lean beef and vinegar.

Osseworst

1 Finely mince the meat and mix together.
2 Extrude into a non-breathable casing, i.e. synthetic.
3 Leave to ferment for 2–4 days.
4 Keep cool at 2–5°C, for a maximum of one week.
5 Eat raw, spread on toast.

Fresh sausage

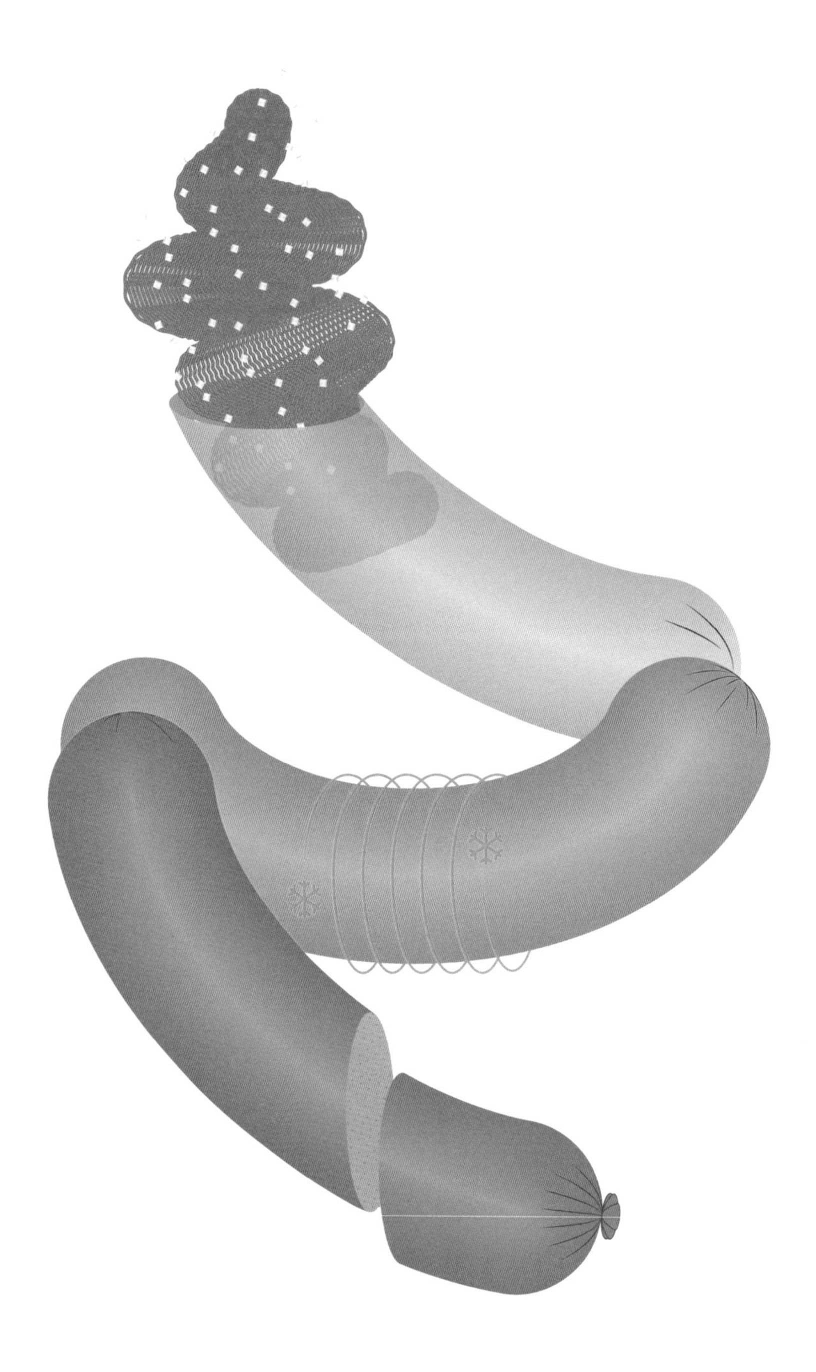

The banger, a type of British sausage, is often enjoyed with a full English breakfast, a dish that also consists of some or all of the following: baked beans, black pudding, grilled tomatoes, mushrooms and scrambled eggs.

Banger

1 Mince the meat to the right coarseness.
2 Extrude the mixture into a breathable edible casing, i.e. intestines or collagen casing.
3 Link into sausage sizes.
4 Store at 2–5°C, for a maximum of 3 days.
5 Grill, fry or roast until cooked through.

Dried sausage

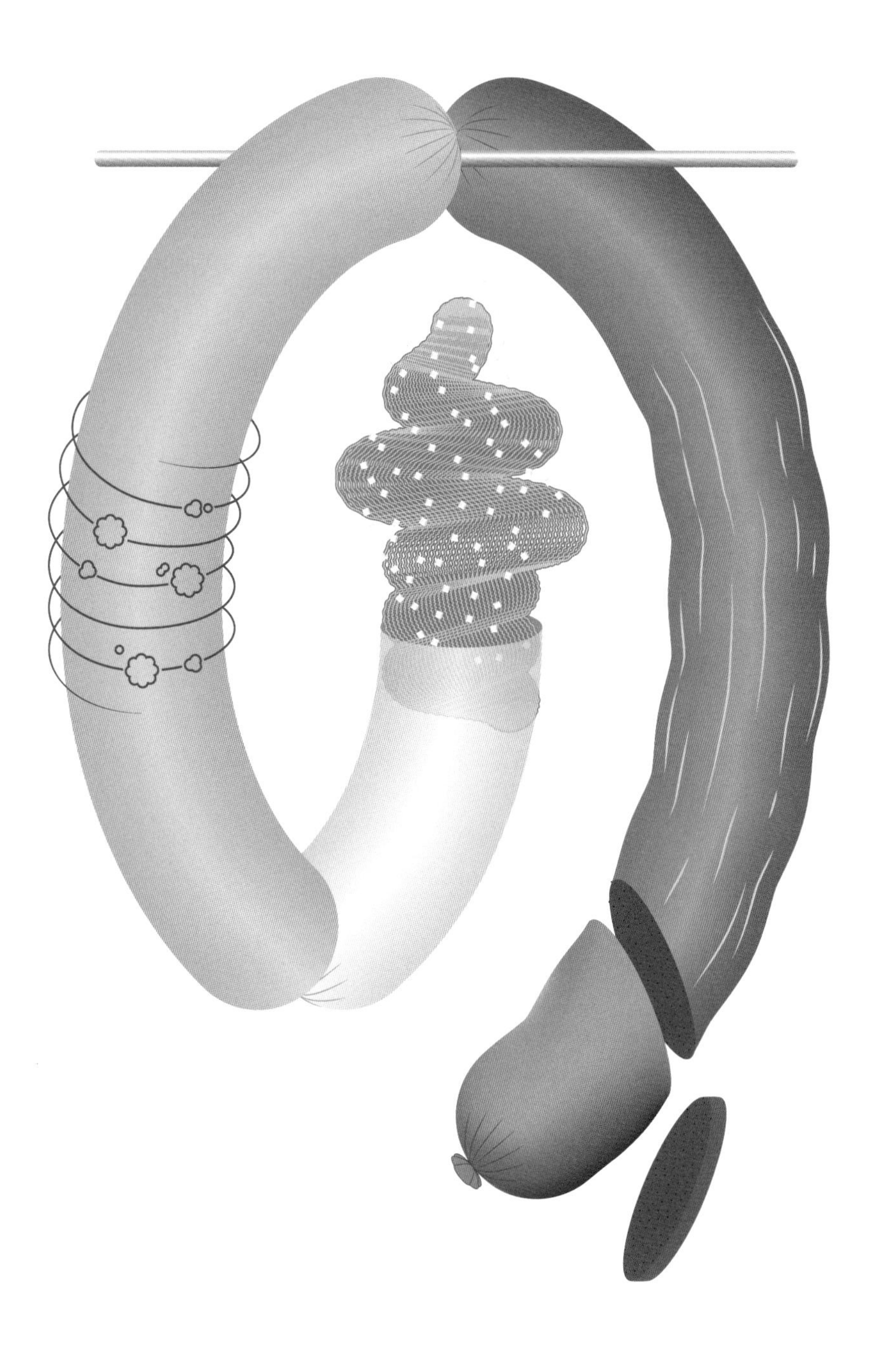

Fuet is a thin type of salami from Catalonia, made from pork. Salami are highly seasoned Italian sausages. They are dried, cured and sometimes smoked.

Fuet

1 Mince the meat finely or coarsely, and mix together with a generous amount of salt.
2 Extrude into a breathable casing, i.e. intestines, or cellulose or collagen casing.
3 Leave to dry in a well-ventilated drying chamber at a temperature of approx. 25°C, for at least 4 days.
4 If desired, dry further, for up to 12 months; this results in a longer shelf life and a dryer, firmer sausage.
5 Remove skin and enjoy sliced or cubed.

Cooked sausage

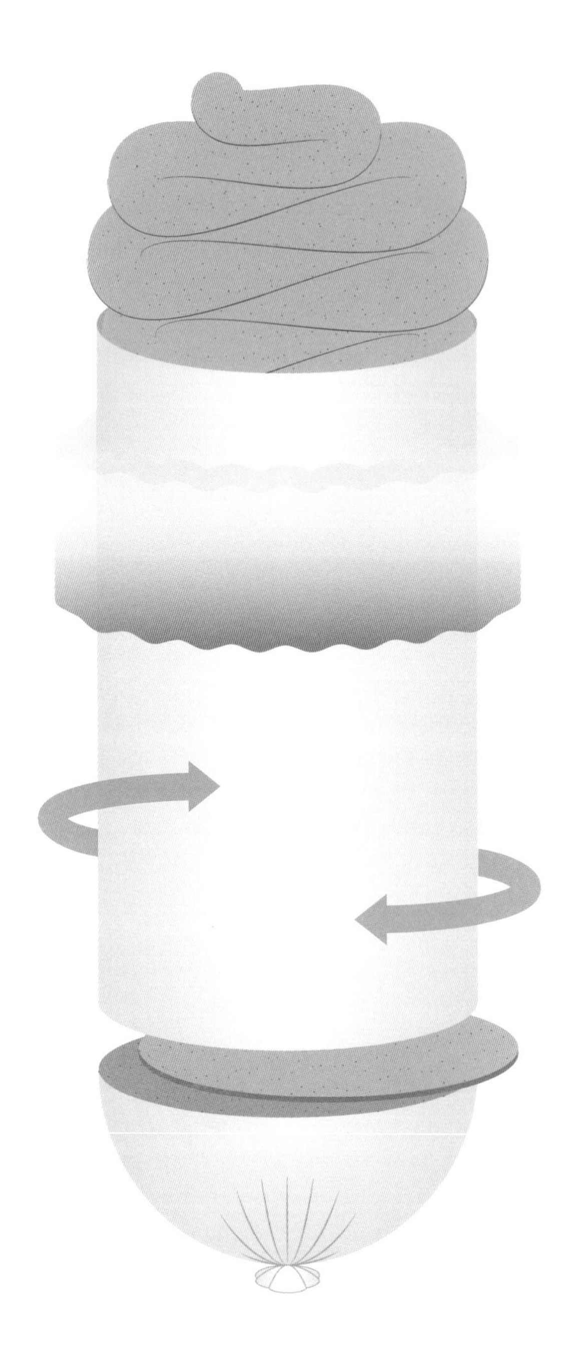

A Dutch variation on mortadella, but without the cubes of fat. The name (*boterhammenworst)* means 'sandwich sausage' and, being mild in flavour, that is what it is commonly used for.

Boterhammenworst

1. Mince and pound the meat into a very fine paste.
2. Extrude into a non-breathable or thick casing, i.e. synthetic or ox runner (cattle intestines).
3. Poach at 80–85°C until the meat is cooked through.
4. Refrigerate at 2–5°C, for approx. 3 months.
5. Eat cold (finely sliced or cubed), or reheat.

Blood sausage

Black pudding, made in a large size and sold in slices to be fried before eating. The flavour of cooked blood is surprisingly mild, and was, in fact, often used in sweet puddings historically.

Black pudding

1. If necessary, mince the other ingredients to the desired size and mix together at a temperature not higher than 45°C.
2. Insert the mixture into a mostly non-breathable and non-edible casing, i.e. synthetic or ox runner.
3. Poach in water at 80–85°C until fully cooked.
4. Before opening, keep at 2–5°C, for 1–3 months.
5. Can be eaten cold, but is mostly enjoyed fried or boiled, and often combined with fruit jams.

Material

Proteins & Protein Chart

The human diet has been identified as the main reason for our species' evolutionary success. The high level of protein in our diet allowed our brain to grow to the size it is now. It is therefore no surprise that we base most of our meals around getting the right amount of protein. What is a protein, why do we need it and where do we find it?

67 From Eximius Forivore to Omnivore
69 Offal
75 Insects
81 Plants
87 Seeds
91 Grains
99 Nuts
103 Legumes

SMALL

Glycine (Gly, G)

Alanine (Ala, A)

NUCLEOPHILIC

Serine (Ser, S)

Threonine (Thr, T)

Cysteine (Cys, C)

HYDROPHOBIC

Valine (Val, V)

Leucine (Leu, L)

Isoleucine (Ile, I)

Methionine (Met, M)

Proline (Pro, P)

AROMATIC

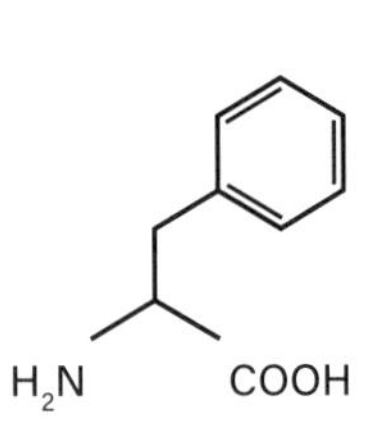

Phenylalanine (Phe, F)

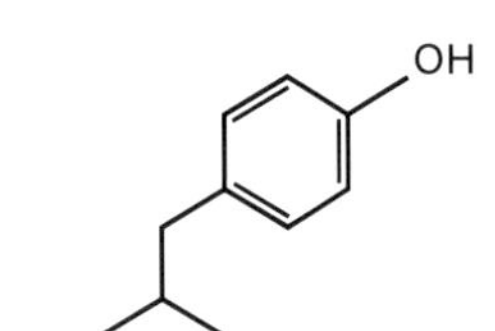

Tyrosine (Tyr, Y)

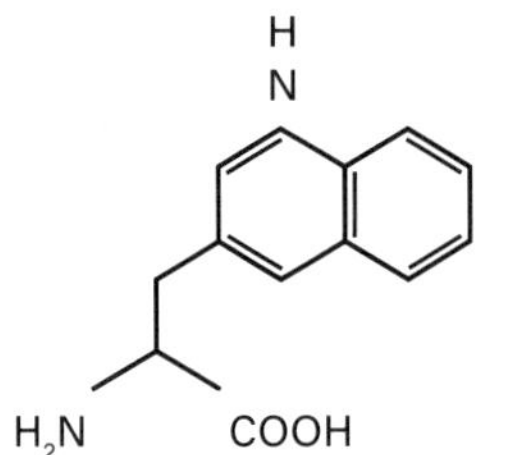

Tryptophan (Trp, W)

ACIDIC

Aspartic Acid (Asp, D)

Glutamic acid (Glu, E)

AMIDE

Asparagine (Asn, N)

Glutamine (Gln, Q)

Histidine (His, H)

Lysine (Lys, K)

Arginine (Arg, R)

Above are all the essential amino acids that together can chain up to become proteins.

Proteins

Nutritional science has undergone a profound revolution in the last few decades. For most of the 20th century, it aimed to define an adequate diet. It determined our body's minimal requirements for chemical building blocks (protein, minerals, fatty acids), for essential cogs in their machinery (vitamins) and for the energy they need to run and maintain themselves from day to day.

Of all the major food molecules, proteins are the most challenging and mercurial. The others – water, fats and carbohydrates – are pretty stable and staid. But expose proteins to a little heat, acid, salt or air, and their behaviour changes drastically. This changeability reflects their biological mission. Carbohydrates and fats are mainly passive forms of stored energy; they are unused structural materials. But proteins are the active machinery of life. They assemble all the molecules that make a cell, including themselves, and tear them down as well; they move molecules from one place in the cell to another; and, in the form of muscle fibres, they move whole animals. They're at the heart of all organic activity, growth and movement. Thus, it is the nature of proteins to be active and sensitive. When we cook foods that contain them, we take advantage of their dynamic nature to make new structures and consistencies. Like starch and cellulose, proteins are large polymers of smaller molecular units. The smaller units are called amino acids. They consist of between ten and forty atoms, mainly carbon, hydrogen and oxygen, with at least 1 nitrogen atom in an amine group NH2, which gives the amino acids their family name. There are about twenty different kinds of *amino acids* that occur in significant quantities in food. Particular protein molecules are dozens to hundreds of amino acids long, and often contain many of the twenty varieties. Short chains of amino acids are called peptides.

Amino acids and peptides contribute flavour to a dish, and there are three aspects of amino acids that are especially important to the cook. First, amino acids participate in the browning reactions that generate flavour at high cooking temperatures. Second, many single amino acids and short peptides have tastes of their own, and in foods where proteins have been partly broken down – aged cheeses, cured hams, soya sauce, and more – these tastes can contribute to the overall flavour. Most palatable amino acids are either sweet or bitter to some degree, and a number of peptides are also bitter. But *glutamic acid*, better known in its concentrated commercial form *MSG (monosodium glutamate)*, and some peptides have a unique taste that is designated by such words as savoury, brothy and *umami* (Japanese for 'delicious'). They lend an added dimension of flavour to foods that are rich in them, including tomatoes and certain seaweeds as well as salt-cured and fermented products. When heated, sulphur-containing amino acids break down and contribute eggy, meaty aroma notes.

Oxidative damage: Our bodies face regular erosion on a chemical level, which is a concern of contemporary nutrition. We breathe in order to allow oxygen to create energy from fats and sugars, thereby maintaining our cells working. But the creation of energy and similar important actions that require oxygen also make free radicals (unstable chemicals that impact on and harm our cells). Their damaging effect tend to stem from oxygen-related responses, and is therefore called oxidative; there are some molecules and enzymes in the body that are responsible for fighting the oxidative effects, called antioxidants. But the more antioxidants the body gets the better it can defend itself, as it turns out, plants are especially abundant in these.

The protein chart on the following spread was used in the construction of the future sausage. It contains nutritional information extracted from a database by the USDA. The list shows different properties of the different ingredients, for example which have complete and enough amino acid chains to replace the proteins in meat, or which contain a high nutritional value that completes our diet. Protein quality is dependent on having all the essential amino acids in the proper proportions. If one or more amino acid is not present in sufficient amounts, the protein is considered incomplete. An amino acid score of 100 or higher indicates a complete or high-quality protein. If the amino acid score is less than 100 the protein is incomplete and has to be substituted in another way. The nutrient completeness score offers a representation of a food's nutritional strengths. A completeness score of between 0 and 100 is a relative indication of how complete the food is with respect to these nutrients. Although few (if any) individual foods provide all the essential nutrients, the completeness score can help construct sausages that are nutritionally balanced and complete.

Protein Chart

Food item (raw)	Protein per 100 g	Calories per 100 g	Amino acid score	Nutrient score
Egg, full	13	143	136	50
Egg, yolk	16	317	146	50
Beef brisket	18	277	101	31
Beef dried	31	153	87	42
Beef heart	17	110	134	58
Beef kidney (veal)	18	112	128	65
Beef liver	20	135	155	67
Beef mince	5	71	67	31
Beef steak	19	247	144	31
Beef sweetbread (veal)	12	236	98	44
Beef tongue	15	224	109	39
Beef top sirloin, lean	22	127	94	41
Pork bacon	3	128	119	14
Pork ear	25	234	27	18
Pork heart	17	118	141	58
Pork kidney	17	100	133	66
Pork liver	21	134	151	71
Pork mince	17	263	146	30
Pork top loin (chop)	21	123	151	39
Chicken	21	172	134	31
Chicken heart	16	153	146	53
Chicken liver	17	116	149	73
Deer venison	22	157	126	47
Duck	11	404	137	20
Elk game	22	172	127	43
Goose	16	371	151	25
Goose liver (fois gras)	11	462	148	34
Lamb	19	195	141	33
Turkey	22	157	142	35
Anchovy	20	131	148	56
Caviar	25	252	146	60
Cod	18	82	148	45
Lobster	19	90	113	47
Oyster	7	68	95	50
Salmon, farmed	20	208	144	39
Salmon, wild	20	142	148	45
Shrimp	20	106	113	56
Grasshopper dried	48	595	NA	NA
Mealworm dried	58	484	NA	NA

Food item (raw)	Protein per 100 g	Calories per 100 g	Amino acid score	Nutrient score
Cow's milk	3	60	85	45
Ewe's milk	6	108	127	42
Goat's milk	4	69	139	44
Camembert	20	300	134	40
Cottage cheese	12	86	158	36
Goat's cheese	22	364	125	36
Gouda	25	356	138	33
Mozzarella	22	300	85	35
Agar	0,5	26	NA	58
Caper	3	23	NA	85
Chicory	2	23	21	91
Dandelion green	1,5	25	NA	88
Kelp	2	43	79	80
Laver	6	35	75	84
Purslane	1	16	65	75
Spirulina	6	26	103	61
Wakame	3	45	27	81
Chia seed	16	490	115	30
Flaxseed	18	534	92	48
Poppy seed	16	490	115	30
Pumpkin seed	19	446	136	24
Sesame seed	18	573	63	52
Sunflower seed	21	584	88	49
Barley	10	350	73	36
Buckwheat	13	343	99	41
Corn	9	365	55	26
Millet	11	378	38	37
Oats	17	246	86	56
Rye	15	335	80	47
Wheat (bread)	11	266	33	45
Wheat (white)	10	364	43	20
Brown rice	8	370	75	32
White rice	7	365	71	17
Wild rice	15	357	84	41
Miso	12	199	44	50
Soya bean	13	122	90	69
Tempeh	19	193	79	44
Tofu, firm	16	145	106	57
Tofu, silk	5	55	106	39

Food item (raw)	Protein per 100 g	Calories per 100 g	Amino acid score	Nutrient score
Almond	22	581	54	42
Brazil nut	14	656	67	32
Cashew	18	553	100	36
Coconut	3	354	87	26
Hazelnut	15	628	55	36
Peanut	26	567	70	43
Pistachio	21	557	109	42
Pine nut	14	673	77	31
Walnut	15	654	55	26
Cocoa powder	20	228	90	56
Azuki bean	20	329	79	54
Black bean	22	341	104	54
Cranberry bean	23	335	104	55
Chickpea	19	364	106	57
Green bean	2	31	88	87
Kidney bean	24	330	104	57
Lentil	26	353	89	57
Lima bean	21	338	95	56
Mungo bean	25	341	95	54
Pea	8	117	84	80
Pigeon pea	22	343	91	55
Asparagus	2	20	93	94
Avocado	2	160	129	48
Beetroot	2	43	71	63
Broccoli	3	34	72	79
Brussels sprout	3	43	61	90
Carrot	1	35	NA	74
Cassava	1	160	52	27
Cauliflower	2	25	103	80
Celery	1	16	52	81
Courgette	1	16	89	86
Eggplant	1	24	67	73
Kale	3	50	92	85
Mushroom	3	22	56	77
Potato	3	264	112	51
Pumpkin	1	26	56	84
Spinach	3	23	119	91
Sweet potato	2	86	82	55
Tomato	1	18	52	80

Highly nutritious chia seeds, close up. In the past the seeds were used as bird feed in the West, and it is only in recent years that they have become appreciated as a food.

From Eximius Forivore to Omnivore

Eximius Forivore *noun*

An animal or person that feeds only on food available in the supermarket.

From Latin *eximius*, meaning exceptional, super, + *forum, fori* n., meaning market + *vorāre*, to eat greedily.

Omnivore *noun*

An animal or person that feeds on food of both animal and vegetable origin, or any type of food indiscriminately.

From Latin *omnis, omnis, omne*, meaning everything, all + *vorāre*, to eat greedily.

Throughout history, the need to make some foods edible through certain processes is illustrated by the importance of the documentation of findings around cooking, namely cookbooks. Many ingredients, such as plants and vegetables, are toxic before they are cooked. However, through the manipulation and cross-breeding of different plant types we have been able to make many foods non-toxic (or substantially less so). For other foods, we have clear cooking rules that are deeply embedded in our culture, for example the correct temperature for killing salmonella when cooking chicken. Cookbooks have saved our lives many times during our wide-ranging exploration of edibles, but we have become unnecessarily dependent on them and trust word-for-word recipes to the point that little creativity seems permissible or is encouraged.

These limitations are also obvious when looking through an average supermarket. It may seem that we have plenty of products to choose from, but given the array of edibles available on earth it seems a limited selection. In the next chapter, we showcase a selection of well- and lesser-known ingredients that illustrate the diversity of our planet. For example, there is a vast variety of offal to choose from, but it is becoming more and more of a specialised butcher product. And what about the ingredients that grow next to our salad but are considered weeds instead of food? The ingredients listed in this chapter, ranging from classic to new, are selected owing to their potential for replacing protein used in sausages traditionally, for their great nutritional composition or because they are ingredients that we could include in our diets but that are not (yet) in our supermarkets.

Each of the following categories contains a group of ingredients chosen for their individual properties. The groups do not attempt to be complete; that would mean defining every single ingredient in a category, which is not our aim here. Rather, we wish to give a sense of the ingredients possible for future sausages.

Ever since humans invented agriculture, we have been trying to optimize crops according to our needs, and we should not stop now. By crossbreeding crops we can make them more efficient, better tasting or even create new flavours; it is part of our evolution and, with the right aim in mind, it can be an exciting form of design.

The following seven sections remind us of forgotten ingredients and inspire us with new ones: be an *omnivore* once more!

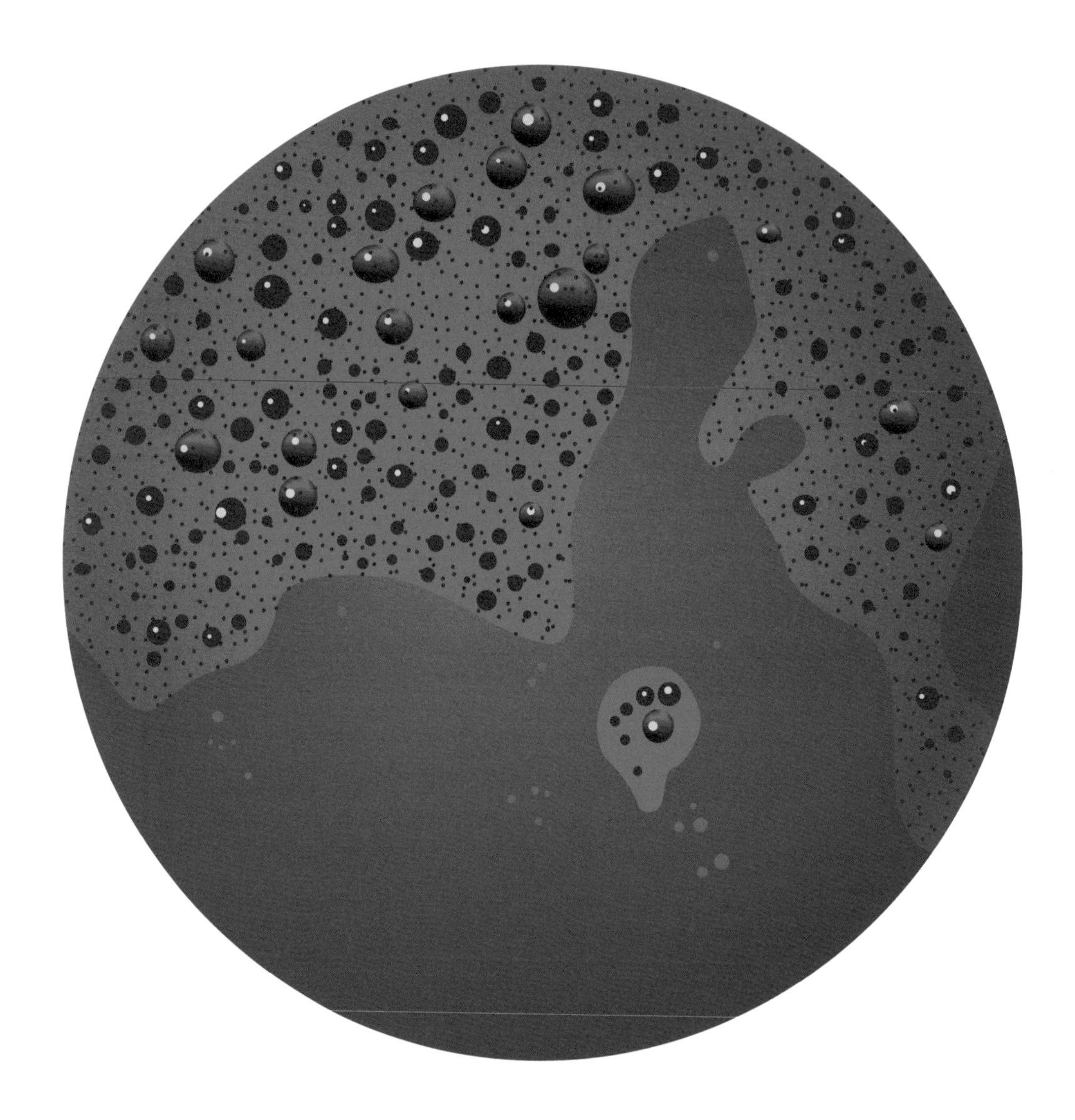

Sanguinaccio dolce, a traditional Italian creamy, rich dessert has a salty, slightly metallic tang to it from pig's blood, which brings out the flavour of the cocoa.

Viscerivore

Viscerivore *noun*

An animal or person that feeds on the entrails and internal organs of animals.

From Latin *viscus, visceris* n., meaning entrails, heart, innermost part of the body, vitals + *vorāre*, to eat greedily.

When mankind began hunting, obtaining an entire animal for slaughter was a prized event; it was unthinkable to let any part of the beast go to waste, from the snout to the intestines to the brain. However now, of all the foods eaten around the world today, none is inherently revolting; their status is entirely based on cultural expectations and conventions. As a result, many people who are perfectly content to be carnivorous are uncomfortable with, and even squeamish about, the notion of eating the guts of an animal. Oddly enough, when asked whether someone likes *pâté*, or even the famous *foie gras* (both of which are made from liver), few people say they dislike it.

The process of rearing, slaughtering, butchering, cooking and, ultimately, eating an animal is completely alien to the increasingly urbanized population in westernized countries. These days, meat is sold in convenient portions, neatly packaged and divorced, as far as possible, from any link to an animal. Eating an organ reminds many people that they are, after all, eating a food that was once a vital part of a living animal. It is unfortunate that this has become an off-putting reality for the majority of consumers.

Both the biological and the culinary differences between various innards are far greater than those between cuts of meat taken from various skeletal muscles.

> *A liver is entirely different from a sweetbread, and a kidney has nothing in common with bone marrow; but a rib steak is not vastly different from a rump steak.*

Cuisines the world over are resplendent with recipes that celebrate various types of offal. Across Asia, Eastern Europe, Africa, the Middle East and Latin America, myriad regional dishes use every part of an animal to delicious ends. Once upon a time, cooked offal dishes were equally popular in Europe and North America, and it is time they regained this status.

The word meat is most commonly used to refer to the limb-moving skeletal muscles of animals. But skeletal muscle only accounts for about half of the animal body. The various other organs and tissues are also nutritious and offer their own diverse, often pronounced, flavours and textures. Moreover, the non-skeletal muscles (stomach, intestines, heart and tongue) generally contain much more connective tissue than ordinary meats – up to three times as much – and thus greatly benefit from slow, moist cooking to dissolve their collagen. Unlike standard meats cut from discrete and largely sterile skeletal muscles, many organ meats carry superfluous matter. They are often trimmed and cleaned before cooking, then blanched, or covered with cold water that is brought to a simmer slowly. The slow heating first washes proteins and microbes off the meat, then coagulates them and floats them to the water surface where they can be skimmed off. Blanching also subdues strong odours on the surface of the meat.

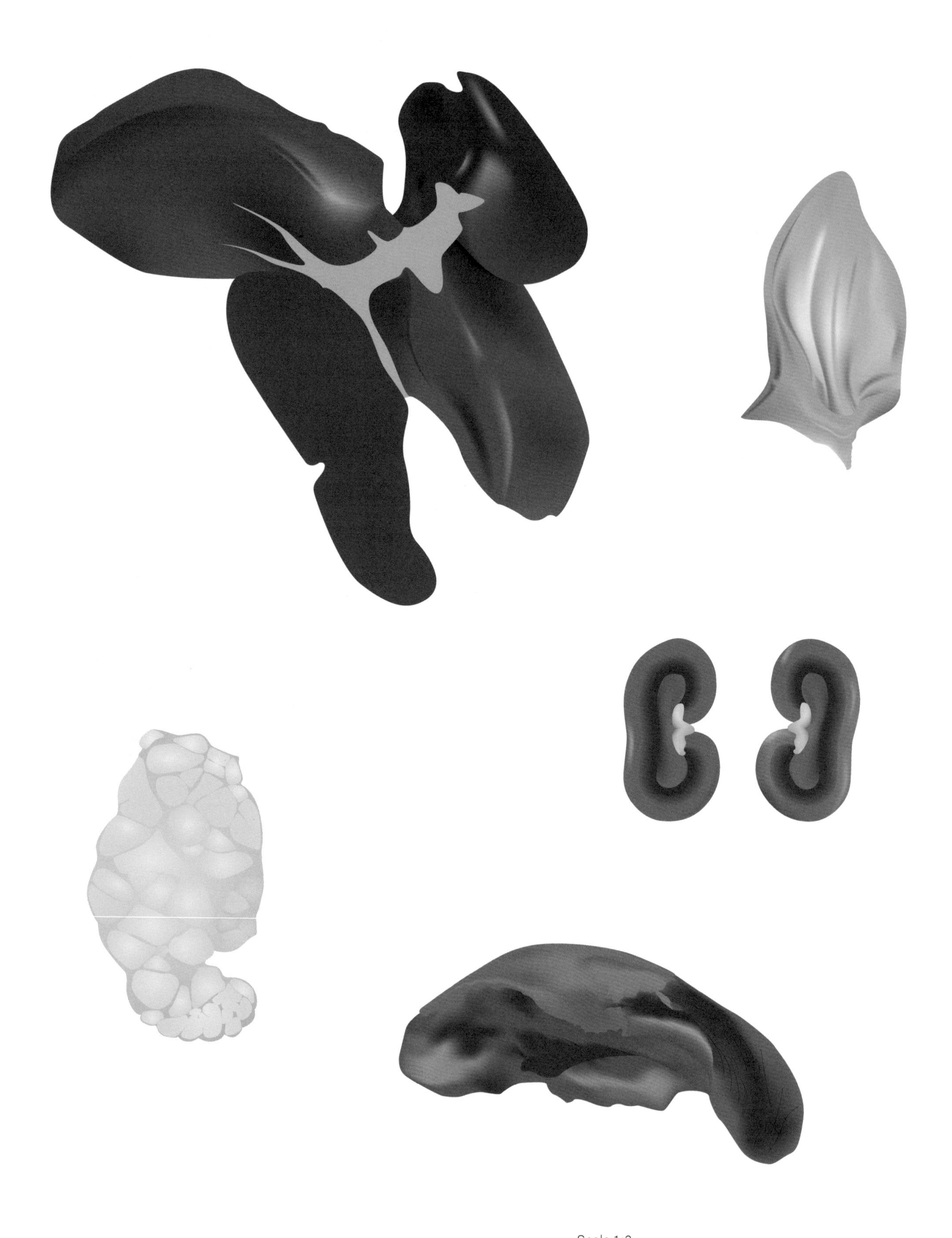

Scale 1:3

Clockwise, from top left, a variety of offal: pork liver, pork ear, pork kidneys, beef tongue, veal sweetbread.

Listed below are some different types of offal:

Blood is perhaps the ingredient that divides the most. Some embrace it as a healthful and inherently natural food, whereas others think of eating blood as an abomination. As a food, it is rich in vitamins and protein and is the best available source of iron. In fact, it is the iron-bearing haemoglobin molecules that give blood its characteristic metallic taste. The thickening and gelling properties of blood make it a popular ingredient worldwide. Many classic sauces are thickened *au sang* (French for 'with blood'). *Civet*, jugged hare and traditionally made *coq au vin* include blood as an ingredient, both for its flavour and to add texture to the sauce. Blood is also combined with bland starches, such as oatmeal and rice, and packed in casings to make the ubiquitous blood sausages found throughout Europe. Blood soup is a somewhat less common Nordic dish, and tofu-like blood cakes are often served in Asia.

The liver is an essential biochemical component of the animal body. Most of the nutrients that the body absorbs from food arrive at the liver first, where they are either stored or processed for distribution to other organs. It is a delicate organ that is at its best when cooked briefly; long cooking simply dries it out. The characteristic flavour of liver has been little investigated, but seems to depend mostly on sulphur compounds which get stronger with prolonged cooking. Generally, both flavour and texture coarsen with age. Liver is mostly eaten in the form of a *pâté*, partly because of its great binding quality. This also makes liver an excellent sausage ingredient.

The skin, cartilage and bones. Cooks do not usually welcome large amounts of tough connective tissue in their cuts of meat. But taken on their own, animal skin, cartilage and bones are valuable exactly because they consist mostly of connective tissue and are therefore full of collagen (skin also provides flavourful fat). Connective tissue has two uses. First, in long-cooked stocks, soups and stews, it dissolves out of bones or skin to provide large quantities of gelatine and a substantial body. Second, it can be turned into a delicious dish itself, with either a succulent gelatinous texture or a crisp, crunchy one, depending on the cut and the cooking method. A long cooking process with plenty of moisture can result in tender veal ears, cheeks and muzzle for *tête de veau*, as well as better beef tendon or fatty pork skin. A briefer cooking process produces crunchy or chewy cartilaginous pigs' ears, snouts and tails, and rapid frying gives crisp pork rinds.

Kidneys are, in mammals, bean-shaped organs that are surrounded by generous layers of fat, which provide both insulation and protective padding. Together, the pair of kidneys in an animal help regulate the composition of blood by filtering it, acting as a reverse osmosis filter for the body effectively. Crucially, the kidneys also keep the volume of water in the body constant by removing excess water from the blood as they carry away waste in the form of urine. Because of their organic role, kidneys spoil quickly, and they must be eaten while very fresh, and certainly within one or two days. The layer of fat surrounding a kidney has a pronounced flavour. Beneath this fat, the kidney itself is sheathed in a thin membrane of tough connective tissue that should be peeled away. At the centre of a kidney is a hardened core that should be carved out before cooking. Searing kidneys quickly over a grill and serving them rare to medium rare is one popular approach. In Britain, broiled kidneys were once a common breakfast dish, but these days braised kidneys in a steak and kidney pie are more popular.

Sweetbreads are blessed with a delicious-sounding (if slightly misleading) name and actually consist of two different glands: the *thymus* and the *pancreas*. The thymus is cylindrical and comes from the throat of the animal, whereas the pancreas is more spherical and is located in the chest near the heart. All mammals have both glands when they are young, but as an animal grows, the thymus shrinks in size, and most of its mass is replaced by fat. Because of their similarities, sweetbreads and all other glands are prepared and cooked in analogous ways. Freshness is essential: a sweetbread should have only a mild odour and be pale pink to light grey in colour. As with the liver, the cells that compose sweetbreads are filled with large amounts of enzymes that can quickly spoil the texture and flavour. Refrigeration, or even freezing, will slow or halt the destructive work of these enzymes. When cooking sweetbreads, heat them to the desired core temperature relatively quickly, and serve them promptly: the cooking temperatures that achieve the ideal soft and creamy texture are not high enough to destroy the enzymes that will ultimately turn them to a chalky mush. The traditional techniques of either quickly sautéing, or alternatively breading or frying sweetbreads, or simply grilling them, all work well. But cooking them *sous vide*, a method of cooking vacuum-sealed food in a water bath at a constant temperature, is ideal. The steady temperature of a water bath makes it easy to achieve a perfectly rich and silky consistency from the surface to the core.

The tongue is technically a skeletal muscle, just like tenderloin. However, unlike the tenderloin, it is not a single muscle but rather an intricately organized network of muscles; each muscle runs in a different direction to give the tongue the dexterity it needs to push food around during chewing. In humans, the assortment of muscles also makes it possible for the tongue to form speech. A tough tissue, closely related to skin, also covers the tongue, the upper surface of which is dotted with various receptors that sense everything from tastes such as saltiness and sweetness to the painful sensation of capsaicin, as well as the temperature and texture of food. The underlying muscle itself is fairly tough and generally benefits from being cooked slowly at low temperatures, making it meltingly tender owing to its high collagen content, which is converted into gelatine.

Heart is the only animal muscle that moves incessantly as long as the animal is alive. Unsurprisingly, this organ is unlike any other muscle. It is closely related to skeletal muscles, such as the tongue, but has adapted to be highly resistant to fatigue. The sheaths of connective tissue that cover cardiac muscle fibres are much stronger than those in skeletal muscles, and they contain a large fraction of elastin, which helps the heart rebound from each contraction to refill with blood. This high elastin content is the source of the distinctive, resilient texture of cooked heart meat. Heart is very lean meat and is edible in virtually all animals. A popular strategy in many parts of the world is to grill thinly sliced or diced heart meat. It should be charred but still rare so that the meat is pleasantly chewy but not tough.

Bones are also considered an organ, although people rarely think of them as such. Anatomically speaking, the skeleton is a cartilaginous substance impregnated with calcareous salts, a mineralized tissue that is porous and slightly flexible but still extremely strong. Besides supporting the body, the bones also supply blood: the soft marrow at the centre of bones creates both red and white blood cells. Joints, which are complex assemblies of cartilage, tendons and ligaments, are crucial ingredients in many stocks and sauces. Although the bones themselves do not contain much flavour, these connective tissues and the pieces of flesh attached to them make ribs, oxtails, shanks, and other bones very flavourful indeed. The red marrow inside smaller bones has little flavour, but the fatty white marrow in large bones such as the *femur* (thigh bone) is delicious.

The stomach and Intestines of mammals have always been prized as containers: the intestines for sausages, and the stomach for Scottish *haggis* and Icelandic *blóðmör* (technically, a type of sausage), in both of which the stomach of a sheep is stuffed with blood, oats and fat. In the culinary world, intestines are often referred to as *chitterlings* or *chitlins*, whereas the stomach is called *tripe* or *maw*.

> Menudo, *a Mexican tripe and hominy (coarsely ground corn) soup, is considered a world-class hangover remedy and is often served on Sundays.*

There exist several varieties of tripe, even from the same animal, because ruminants such as goats and cows have multiple stomachs. Each stomach performs a different function and thus has a lining of slightly different composition. The most commonly seen are the *rumen*, or smooth tripe, and the *reticulum*, or honeycomb tripe. Steamed and then grilled or fried, these tissues can have a wonderful texture: crispy, chewy and tender all at once. When cooked sous vide for long enough or pressure- cooked briefly, they can also be transformed from tough and chewy to soft and tender, with a unique texture revered in many parts of the world.

Fat is perhaps the only type of offal that we consider a culinary staple. Sections that contain a lot of fat, such as the fatty areas around organs and in the backs of pigs, are the most prized. At first glance, these fats may seem accessible and ready to use, but in fact the fat still needs to be removed from the cellular structure and rendered before it will combine with other ingredients. Animal fat is primarily used as a cooking medium or ingredient, rather than as a food in its own right. Some fats, such as the web-like caul fat that surrounds the abdominal organs in some animals, are strong enough that they are used as natural casings for sausages or pâtés.

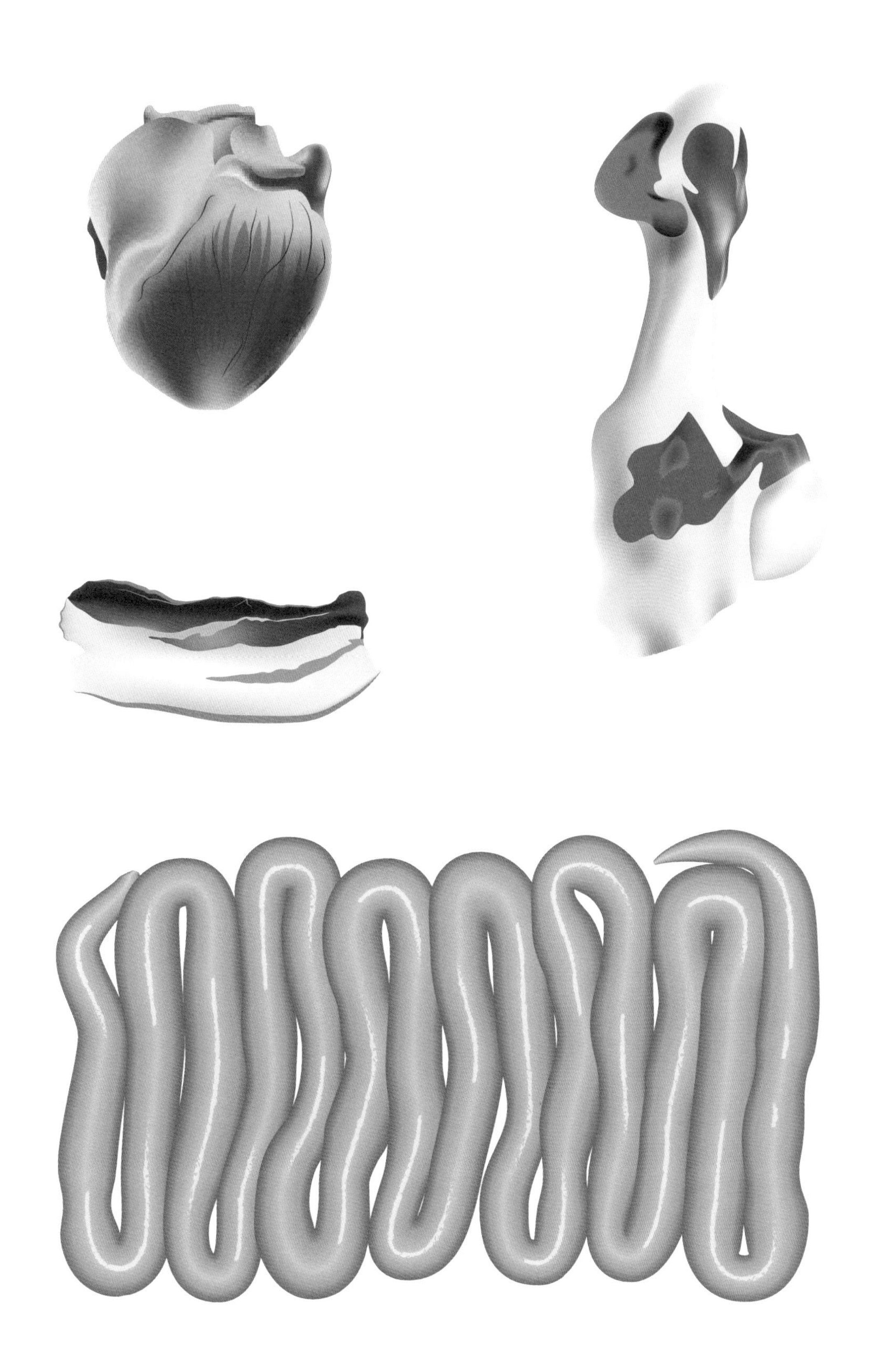

Scale 1:3

Clockwise, from top left, a variety of pork offal: heart, femur, intestines, belly fat.

Grasshoppers (here a *lophacris cristata*), or *chapulines*, are said to have been a Mexican staple for so long that the inhabitants of Oaxaca, an area famous for having chapulines as part of its menu, called shrimp sea-crickets at first encounter.

Insectivore

Insectivore *noun*

An animal or person that feeds on insects, worms, and other invertebrates.

From Latin *insectum, insecti* n., segmented (animal) + *vorāre*, to eat greedily.

If we are to believe the USDA's book *Edible Insects: Future Prospects for Food and Feed Security* (2013), our future source of food is insects. There are over one million different types of insects on the planet, and 1,900 of these are considered edible. Moreover, around 80 per cent of the world's inhabitants currently consume insects. The official term for eating insects is *entomophagy*, from Greek *éntomon*, insect + *phagein*, to eat. Culture affects our approach to food constantly, which in turn has been influenced by religious faith. There are references to the consumption of insects (known as entomophagy) in the scriptures of the three major world faiths. Desert locusts are mentioned as food in the book of Leviticus, and entomophagy appears in Jewish literature. According to Professor Zohar Amar, of Bar-Ilan University, in Israel, kosher locusts were eaten widely historically, but as an understating of which types of insects were allowed in the *Torah* waned among the Jewish diaspora and with westernization, so did the practice. In fact, it is only with Yemeni Jews that the practice remained.

The history of entomophagy is well documented by Shimon Fritz Bodenheimer in his book *Insects as Human Food* (1951). In the Middle East, as far back as the 8th century BCE, servants were thought to have carried locusts arranged on sticks to royal banquets in the palace of Ashurbanipal, king of Assyria, in Nineveh. The first reference to entomophagy in Europe was in Greece, where eating cicadas was considered a delicacy. Aristotle (384–22 BCE) wrote the following in his *History of Animals*: 'The larva of the cicada on attaining full size in the ground becomes a nymph; then it tastes best, before the husk is broken, i.e. before the last moult.' He also mentioned that, of the adults, females taste best after copulation because they are full of eggs.

References to entomophagy continued throughout the Mediterranean region and the centuries. In the 2nd century BCE, Diodorus of Sicily called people from Ethiopia *Acridophagi*, or 'eaters of locusts and grasshopper'. Similarly, in Ancient Rome, author, natural philosopher and naturalist Pliny the Elder – author of the *Historia Naturalis encyclopaedia* (77–79) – spoke of *cossus*, a dish highly coveted by Romans. According to Bodenheimer, cossus is the larva of the longhorn beetle *Cerambyx cerdo*, which lives on oak trees. Literature from ancient China also cites the practice of entomophagy. Li Shizhen's *Compendium of Materia Medica*, one of the largest and most comprehensive books on Chinese medicine of Ming Dynasty China (1368–1644), displays an impressive record of all foods, including a large number of insects. The compendium also highlights the medicinal benefits of the insects. China (1368–1644) displays an impressive record of all foods, including a large number of insects. The compendium also highlights the medicinal benefits of the insects.

Modern-day entomophagy

The Italian entomologist and naturalist Ulisse Aldrovandi, born in 1522, is considered the founder of the modern-day study of insects. His *De Animalibus Insectis Libri Septem* (1602) is rich in references and concepts derived from his studies as well as original observations. Aldrovandi, a specialist in cicadas, suggested that insects were important food items in ancient Far Eastern civilizations, namely China, as far back as several centuries BCE. Yet it was not until the 19th century, when explorers brought back observations from tropical countries, that the Western world grew familiar with the practice of entomophagy. Explorers' accounts of Africa, such as those of David Livingstone and Henry Morton Stanley, which featured stories of insect eating, were instrumental in introducing the practice to the West. In 1857, German explorer Barth Heinrich, for example, wrote in his book *Travels and Discoveries in North and Central Africa* that people who ate insects 'enjoy not only the agreeable flavour of the dish, but also take a pleasant revenge on the ravagers of their fields', an interesting take on agricultural pests.

Scale 1:1

Clockwise, from top left, a variety of insects: bee larva, cockroach, silkworm, mealworm, witchetty grub.

In the USA, swarms of Rocky Mountain locusts (*Melanoplus spretus*) regularly swept across the western half of the country (as far north as Canada) in the 19th century, devastating farming communities, as described in Jeff Lockwood's *The Management of Insects in Recreation and Tourism* (2004). One famed sighting estimated that the locusts spanned more than 500,000 square kilometres. This swarm weighed an estimated 27.5 million tonnes and consisted of some 12.5 trillion insects, which according to Guinness World Records was the greatest concentration of animals ever documented.

The first State Entomologist for the state of Missouri, Charles Valentine Riley, considered the far-reaching 1873-77 Rocky Mountain locust plague, suggesting consuming the insects to restrict their numbers:

'Whenever the occasion presented I partook of locusts prepared in different ways, and one day, ate of no other kind of food, and must have consumed, in one form and another, the substance of several thousand half-grown locusts. Commencing the experiments with some misgivings, and fully expecting to have to overcome disagreeable flavour, I was soon most agreeably surprised to find that the insects were quite palatable, in whatever way prepared. The flavour of the raw locust is most strong and disagreeable, but that of the cooked insects is agreeable, and sufficiently mild to be easily neutralized by anything with which they may be mixed, and to admit of easy disguise, according to taste or fancy. But the great point I would make in their favour is that they need no elaborate preparation or seasoning.'

However, the booklet *Why Not Eat Insects*? (1855) by British entomologist V.M. Holt is considered to have alerted the most people to the issue, proposing that 'one of the constant questions of the day is, "How can the farmer most successfully battle with insects devouring his crops?" I suggest that these insects should be collected by the poor as food. Why not?'

As a Victorian, Holt believed in the moral values of his day, including caring for resources and for the poor, and wondered why people were happy to eat crustacea but not the related insects. He also made reference to the eating of insects in other cultures:

'If I bring forward examples from ancient times, or from among those nations, in modern times, which are called uncivilized, I foresee that I shall be met with the argument, "Why should we imitate these uncivilized races?" But upon examination it will be found that, though uncivilized, most of these peoples are more particular as to the fitness of their food than we are, and look on us with far greater horror for using, as food, the unclean pig or the raw oyster, than we do upon them for relishing a properly cooked dish of clean-feeding locusts or palm-grubs.'

Holt was a visionary in terms of entomophagy, and while his ideas did not filter into the cuisine of his day, perhaps the time is ripe for his ideas now.

Humans rely on insects to fulfil myriad ecological acts and to provide some medical remedies (maggot therapy) and products (silk and honey), as well as in biocontrol of certain species of pest, pollination and waste bioconversion to aid the quality of soil. Insects have also captured the human imagination, in a variety of art forms and practices. When considered worldwide, the most widely consumed groups are beetles, caterpillars, bees, wasps, ants, grasshoppers, locusts, crickets, cicadas, leaf- and plant-hoppers, scale insects and true bugs, termites, dragonflies and flies.

Since there are many differing points of view when it comes to explaining the taste or preparation of insects, combined with little fact-based information, the details given here are slightly shorter than for the other ingredients in the book, and should be taken more as guidelines than given facts.

Listed below a selection of edible insects:

From a nutritional standpoint, **bee larvae** (*Apidae*) eaten as they were historically is a complete food. This method, used by beekeepers by way of disposing of the larvae during honey production, includes eating the honey, larvae and pupae while still inside the comb. The larvae of honeybees are said to be similar to nuts or bacon or nuts, and they are consumed regularly and in a variety of ways – raw, stewed, fried – in Mexico, South America, Africa, Australia and Africa. Because the idea of honeybee larvae is already present in our food culture, this might make it one of the gateway insects towards getting the Western world excited about entomophagy.

Cockroaches fall under the *Blattodea* order, the same order as termites do. Formerly, the termites were considered a separate order, *Isoptera*, but genetic and molecular evidence suggests an intimate relationship with the cockroaches, both cockroaches and termites having evolved from a common ancestor. Cockroaches originate in Thailand and Cambodia. When cooked they are often blanched or fried before eating, with the legs, wings and carapace removed. The taste of the cockroach is most often compared to chicken.

Silkworms (*Bombyx mori*) originate in Korea, Japan and China. They can be fried as a snack or boiled with seasoning. Boiled silkworms have a pungent, almost bitter smell and a similar taste. When you bite down on them, they pop juicily in your mouth. The taste might be an acquired one at first, but can certainly become an appreciated one since silkworms are considered a delicacy and are sometimes compared to seafood.

Mealworms (*Tenebrio molitor*) are one of the few edible insects to have strongly penetrated Western food cultures, partly due to their easy rearing, for example on fresh oats, wheat bran or grain, with sliced potato, carrots or apple as a moisture source, possibly also because mealworms originated in Europe. Now largely available worldwide, they are often baked or fried and marketed as a healthful snack food. Their taste has notes of shrimp and nuts.

Witchetty grub (*Endoxyla leucomochla*) refers to the large, white, wood-eating larvae of several moths (*Cossidae* and *Hepialidae*) and beetles (*Cerambycidae*) found in Australia. However, the term applies mostly to the larvae of the cossid moth, which can be found 60 centimetres below ground feeding on the roots of river red gum trees. The grub is the most important insect food of the desert and was a staple in the diets of Aboriginal women and children. The grubs can be eaten either raw or lightly cooked in hot ashes, and they are sought by Aborigines as a high-protein, high-fat food. The raw witchetty grub tastes like almonds; when cooked, the skin becomes crisp like that of roast chicken and the inside turns light yellow in colour.

Grasshoppers, locusts and crickets are insects of the suborder *Caelifera* within the order *Orthoptera*. There is no taxonomic distinction is made between locust and grasshopper species but locusts are certain species of short-horned grasshoppers in the family *Acrididae* that have a swarming phase. This means that though these insects are usually solitary, under certain circumstances they become more abundant and change their behaviour and habits, becoming gregarious. Grasshoppers and crickets are considered a delicacy and are eaten in many African, Middle Eastern and Asian countries. They can be cooked in many ways but are often fried, smoked or dried. The wings and legs are often removed. Because of the many different varieties, the taste can range from seafood to mushrooms to nuts, and many more.

Globally, there are some 12,500 species of **ant** classified by humans and from those there are many different types consumed by humans. These range from *Formicidae* (ants) such as the sweet tasting black honey ant (*Camponotus inflatus*) by Aborigines in Australia, or fried *hormigas culonas* (large-bottomed ants, or *Atta laevigata*) eaten as a snack in cinemas in Colombia. An ant's defence system dictates its flavour: mint, kaffir lime, lemongrass or orange blossom. The sour flavour of ants comes from the naturally occurring formic acid, which was identified in Formicidae first.

Dragonflies (*Anisoptera*) are an edible insect encountered less often in writing, but they are most definitely eaten. Originating in Bali and China, dragonflies are often caught with a net and mostly grilled or fried, eaten with the wings and legs removed. They allegedly taste appealingly like seafood.

There are forty-three **Termite** (*Termitoidae*) species used as food by humans or fed to livestock. These insects are particularly important in less developed countries where malnutrition is common, as the protein from termites can help improve the human diet. Termites are consumed in many regions globally, but the practice has only become popular in developed nations in recent years. In Africa, *alates* (the winged form of social insects) are an important factor in the diets of native populations. Tribes have different ways of collecting or cultivating insects; sometimes they will collect soldiers from several species. Though harder to acquire, queens are regarded as a delicacy. Termite alates are high in nutrition, with adequate levels of fat and protein. They are considered pleasant in taste, having a nut-like flavour after being cooked.

Cicadas (*Cicadoidea*) are a type of beetle, originally from Japan and Thailand. They are often eaten blanched or fried, with the legs, wings and carapace removed. Cicadas have taste notes reminiscent of asparagus. They were eaten in ancient Greece, and are consumed today in China, Malaysia, Burma, Latin America and the Congo, both as adults and (more often) as nymphs. Female cicadas are prized for being meatier. In 2011, cicadas were incorporated into a single batch of ice cream in Columbia, Missouri, at a Sparky's branch. Unfortunately, the ice-cream parlour was advised by the public health department against making a second batch.

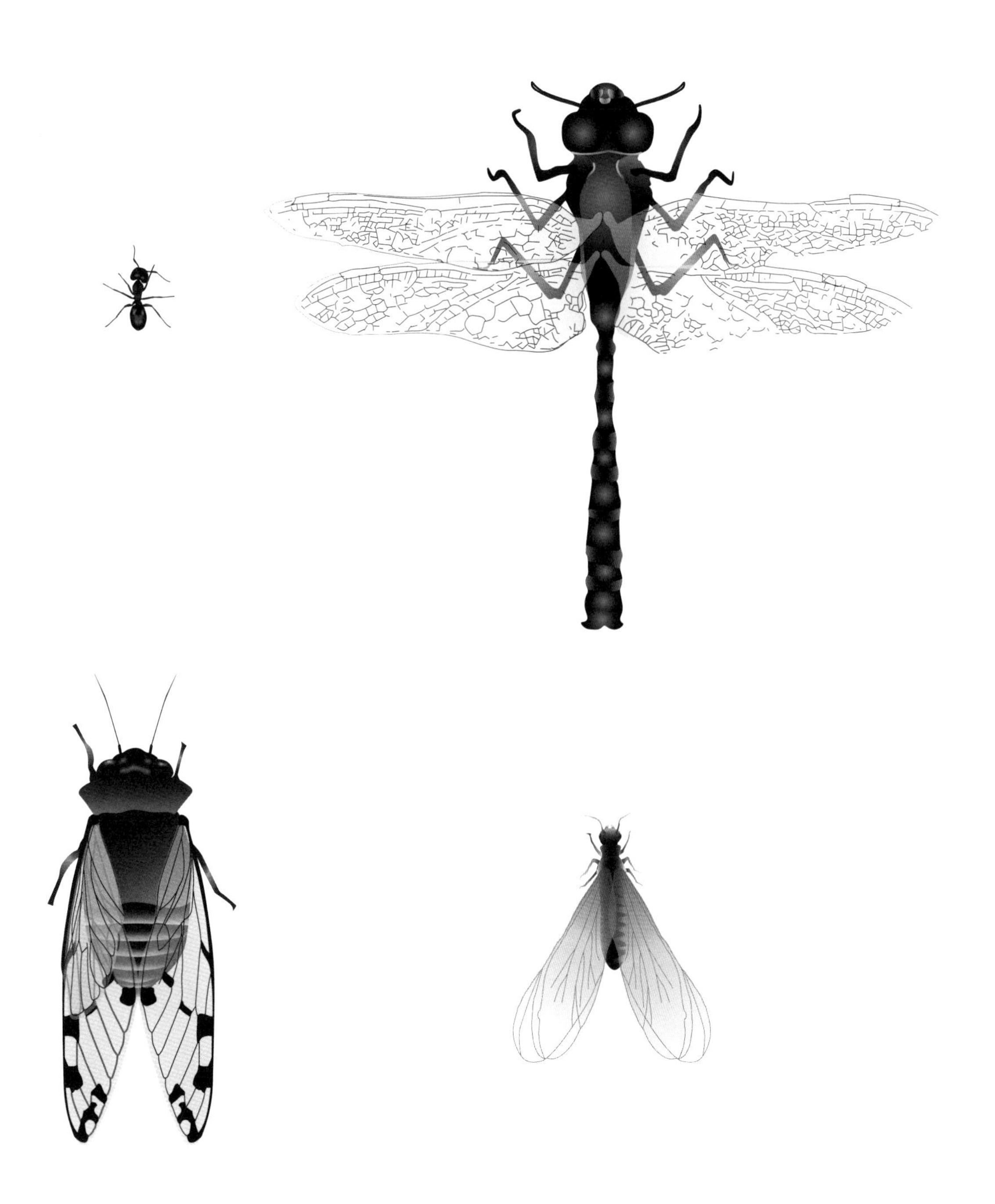

Scale 1:1

Clockwise, from top left, a variety of insects:
black worker ant, dragonfly, termite alate, cicada.

Because of the buoyancy in water, seaweeds do not need a predominantly stalky construction to defy gravity, as plants do. Seaweed leaves can therefore grow much bigger than would be possible above water.

Herbivore & Fucivore

Herbivore *noun*

An animal or person that feeds on plants.

From Latin *herba, herbae* f., herb + *vorāre*, to eat greedily.

Fucivore *noun*

An animal or person that feeds on seaweed and kelp.

From *fucus, fuci* m., bee-glue, propolis, dye, seaweed + *vorāre*, to eat greedily.

Human beings have always been plant eaters. For a million years and more, our omnivorous ancestors foraged and lived on a wide range of wild fruits, leaves and seeds. Around 10,000 years ago, they began to domesticate a few grains, seed legumes and tubers, all of which are among the richest sources of energy and protein in the plant world, and can be grown and stored in large quantities. This control over the food supply made it possible for many people to be fed reliably from a small patch of land. Hence, cultivation of the fields led to settlement, the first cities and cultivation of the human mind. On the other hand, agriculture reduced drastically the variety of plant foods in the human diet. Only recently have we begun to understand how the human body still depends for its long-term health on a varied diet rich in fruits and vegetables, herbs and spices. Happily, modern technologies now give us unprecedented access to the world's large variety of edible plants. The time is ripe to explore this fascinating and still evolving legacy of natural and human inventiveness.

The plant world encompasses earthy roots, bitter and pungent and refreshing leaves, perfumed flowers, mouth-filling fruits, nutty seeds, sweetness and tartness and a stringency and pleasing pain, and aromas by the thousands. It turns out that this exuberantly diverse world was born of simple, harsh necessity. Plants cannot move as animals do. To survive their immobile, exposed condition, they became virtuosic chemists. They construct themselves from the simplest materials of the earth itself – water, rock, air and light – and thus transform the earth into food on which all animal life depends. Plants deter enemies and attract friends with colours, tastes and scents – all chemical inventions that have shaped our ideas of beauty and deliciousness. And they protect themselves from the common chemical stresses of living with substances that protect us as well. Thus, when we eat vegetables, fruits, grains and spices, we eat the foods that made us possible, and that in turn opens our lives to a kaleidoscopic world of sensation and delight.

Plants are made of gas. Just two things contribute to a plant's weight. A huge fraction is water drawn up by the roots of the plant. Most of a plant's dry matter comes from the air, more specifically from carbon dioxide absorbed by the leaves through intake ports called *stomata*. Hence, fertilizer companies' claims that plants need rich nutrients from soil are true but misleading. For soil nutrients are like vitamins: necessary for good health but only a small contributor to plant mass, which is primarily carbon, oxygen and hydrogen. Thus, it is merely a tiny percentage of the dry weight of a plant that comes from the soil.

Flowers are plant organs that attract pollinating animals with a strong scent, bright colours or both. They therefore can add both aromatic and visual appeal to our foods. But the most important edible flowers in the West are neither colourful nor flowery. Broccoli and cauliflower are immature flower structures, and artichokes are eaten before they have a chance to open. Aromatic flowers have played a more prominent role in the Middle East and Asia. In the Middle East, the distilled essence of indigenous roses, and later of bitter orange flowers from China, has long been used to embellish the flavours of many dishes: rose water in *baklava* and *Turkish delight*, for example, and orange flower water in Moroccan salads and stews and in Turkish coffee. Food historian Charles Perry has called these extracts 'the vanilla of the Middle East'. They were also commonly used in the West until vanilla displaced them in the middle of the 19th century.

There are many plants and flowers that are edible but are not part of our diets anymore. Indeed, historically these ingredients were sometimes incorporated into sausages, for example in the traditional *chorizo verde*: a central American sausage containing a large variety of greens such as spinach, coriander, parsley and often *tomatillos* (green tomatoes). The reason we went from a diverse and flowerful diet to a much more restricted supermarket-available diet is firstly because, with the industrial revolution and many people's movement from farm to factory, the availability of plants and flowers was reduced. As consumers nowadays rely mostly on supermarkets, shops and farmers' markets, they do not come across many edible plants when shopping, and therefore they are not being eaten. However, in the past decade, blogs and other online sources about foraging and eating directly from nature, as well as high-end restaurants that use foraged food, have sprung up around the world, allowing people to consume a greater variety of ingredients, and contributing to a considerably healthier diet than we have had for a century or so.

Listed below are some edible plants from the northern hemisphere:

Green Algae is a thin, translucent seaweed with an intense bright-green colour. It is traditionally used for human consumption in Asian countries and more recently in Europe. The sea lettuce has a texture like cellophane, which becomes crispy if well roasted. In dishes it is commonly found in very finely chopped salads, as a second skin on grilled fish or crushed into sauces.

Violet flowers (*Viola odorata*), when newly opened, may be used to decorate salads or in stuffing for poultry or fish. *Soufflés*, creams and desserts can be flavoured with violet essence. The young leaves are edible raw or cooked as a somewhat bland leaf vegetable. The flowers and leaves of the cultivar Rebecca, of the Violaceae family of violets, have a distinct vanilla flavour with hints of wintergreen. The pungent perfume of some varieties of *V. odorata* adds inimitable sweetness to desserts, fruit salads and teas, while the mild pea flavour of *V. tricolour* combines equally well with sweet or savoury foods, including grilled meats and steamed vegetables. The heart-shaped leaves of V. odorata provide a free source of greens throughout a long growing season.

Kelps are large seaweeds belonging to the brown algae. *Alginate*, a kelp-derived carbohydrate, is a modern ingredient used as a gelling agent. *Kombu*, a Pacific variety of kelp, is a key ingredient in Japanese cuisine.

Lavender (*Lavandula dentate*) takes its name from the Latin for 'wash' and is familiar from its application in candles and soaps, but the herby and floral notes of this bush of the Mediterranean have lent themselves to use in food since antiquity as well. In their dried form, its blossoms are included in the herb mix: *herbes de Provence* (with marjoram, thyme, basil, fennel and rosemary). While Spanish lavender (*L. stoechas*) brings to mind Indian chutney, the blossoms of English lavender (*L. angustifolia*) are infused in sweets or sauces, or as a subtle garnish.

Nasturtiums (*Tropaeolum majus*) is a flowering plant in the *Tropaeolaceae* family, originating in the Andes – from Bolivia north to Colombia. Garden nasturtiums are grown for their bright coloured flowers but also for their edible leaves and flowers that can be used in salads, imparting a delicately peppery flavour. The seeds are also edible and can be used in the same way as capers. Nasturtiums are related to rocket, cress, mustard greens and Ethiopian mustard rocket. These cresses, which are small, weedy cabbage relatives, come from the Mediterranean region and are especially pungent, with a full, almost meaty flavour. They are often used in salads, to accompany meat or fish or as pizza toppings.

Dandelion greens (*Taraxicum officinale*) can grow continually; if its taproot is left undisturbed it will give a rosette of leaves repeatedly. The leaves are usually made less bitter by blanching. While they grow across the globe naturally, Eurasia is the origin of the majority of farmed types. Furthermore, they have been collected from the wild for millennia, and are sometimes cultivated in small areas.

Wood sorrel (*Oxalis acetosella*) is a variety of sorrel, the startlingly sour leaf of several European relatives of rhubarb and buckwheat that are rich in oxalic acid. Cooks use them mainly as a source of acidity, and they also provide a more generic grassy aroma. Sorrel readily disintegrates with a little cooking into a sauce-like puree that complements fish, but whose chlorophyll turns drab olive from the acidity. The colour can be brightened by pureeing some raw sorrel and adding it to the sauce just before serving.

Angelica (*Angelica archangelica*) grows on grassland in northern Europe and combines the smell of the sweet compound *angelica lactone* with the clean scent of citrus and pine. While contemporary use includes sweets, perfumes, vermouths, liqueurs and gins, the candied stems favoured for centuries from the Middle Ages are no longer made.

Chicories (*Cichorium intybus*) still have the original intense bitterness of lettuce, which has been bred out of most cultivated forms. A number of close lettuce relatives from the genus Cichorium are cultivated and included in salads or cooked on their own, especially to provide a civilized dose of bitterness. Cichoriums include endive, escarole, chicory and radicchio. Growers go to much trouble to control their bitterness. Open rosettes of escaroles and endives are often tied into an artificial head to keep the inner leaves in the dark and relatively mild. The popular Belgian endive, also known as *witlof* (meaning 'white head'), is a double-grown, slightly bitter version of an otherwise extremely bitter chicory. The plant is grown from seed in the spring, defoliated and dug up in the autumn, and the taproot

Scale 1:1

Clockwise, from top left, a variety of plants: wild violet, kelp, lavender, nasturtium, dandelion

Scale 1:1

Clockwise, from top left, a variety of plants: wood sorrel, angelica, chicory, nettle, caper. Opposite page: purslane.

with its nutrient reserves kept in cold storage. The root is then either replanted indoors and kept covered with soil and sand as it leafs out, or else it is grown hydroponically in the dark. The root takes approximately a month to develop a fist-sized head of white to pale-green leaves, with a delicate flavour and crunchy yet tender texture. This delicacy is easily lost, for example, by exposing the heads to light in the market will induce greening and bitterness in the outer leaves, and the flavour to become harsh.

Nettles (*Urtica dioica*) is widespread in the northern hemisphere, but originates in Eurasian. The Urtica dioica weed is known for its irritant sting, a result of the chemicals (histamine, for example) supplied by glands on its leaves. The leaves are covered in hairs whose delicate silicate tips contain the glands. However, the effect can be undone by blanching the leaves in boiling water, thereby releasing and diluting the chemicals, and the blanched leaves can be used for soup, as well as a pasta filling with cheese or stewed.

Capers (*Capparis spinosa*) have been farmed for some two hundred years, yet the wild variety was picked and preserved for millennia. They are the closed flower buds of a Mediterranean bush related remotely to the cabbage family, and, when eaten raw, have the same pungent sulphur compounds.

The salty and sour flavour that capers lend to dishes, including fish, comes from their preservation, which can take a number of forms, vinegar, brine, dry-salt, the latter of which results in raspberry and violet bouquets in place of the characteristic onion and radish.

Purslane (*Portulaca oleracea*) originated in Europe but has reached around the globe. In the 19th century, it was said by Englishman William Cobbett to be worthy of the French and of pigs alone; perhaps coincidentally, this led to the byname pigweed. However, whether used in either vegetarian or meat dishes towards the end of cooking or added fresh to salads, purslane is loved in numerous places for its sharpness. Structurally, it is low, and its leaves – compact and thick – sit on broad stems. The weed, which abounds in abandoned areas and enjoys hot weather, is nowadays farmed in several varieties of larger leaves in pink and yellow. It is rich in nutrients and vitamins such as linolenic acid (an omega-3 fatty acid) and calcium.

Seaweed or algae, is a biological watery group nearly a billion years old, the predecessor to all land plants, and people have eaten hundreds of the more than 20,000 species of algae. They have been central to the diets of the inhabitants of less plant-fertile islands such as Iceland and Hawaii, as well as in the British Isles and the coasts of Asia. In Ireland seaweed is used as a dessert thickener and mashed up in porridge, while in China it is used as a vegetable, and in Japan in salads, soups and as rice wrapping. Seaweed is responsible for the particular smell of the sea, while its taste is clean, deep and briny. Nutritionally, the dry variety is protein rich, and otherwise the majority of seaweed provides high levels of iodine and vitamins A, B, C and E, while from a sustainability standpoint it is replenished within a year or two, is plentiful and keeps for a long time when dried.

> *Further, sushi-wrapping seaweed, which has been grown in Japan for four centuries, monetarily exceeds the entirety of other aqua-cultural products, shellfish and fish included, in terms of farming.*

The watery home of seaweeds has shaped their nature in several ways; for example the buoyancy in water has allowed free-floating algae to minimize tough structural supports and maximize photosynthetic tissue. Some algae (e.g. *nori* and sea lettuce) are essentially all leaf and just one or two cells thick. They are exceptionally tender and delicate. Further, their immersion in salt water of varying concentration has led algae to accumulate various molecules to keep their cells in osmotic balance. Some of these molecules contribute to their characteristic taste. Some examples include the savoury glutamic acid, the sugar alcohol mannitol (which sweet taste is low in calories as it is not metabolizable).

The many physical stresses of ocean life have encouraged some seaweeds to fill their cell walls with large quantities of jelly-like material that gives their tissues strength and flexibility and can help keep coastal species moist when they are exposed to the air at low tide. These special carbohydrates turn out to be useful for making jellies (agar) and for thickening various foods (algin, carrageenan).

The opium poppy is the only species of *Papaveraceae* that is grown as an agricultural crop on a large scale. Above: anatomical drawings of the poppy. Left: the flower. From top: cross-section, top-view section and side view of the papaver (flower bud), which contains the seeds.

Seminivore

Seminivore *noun*

An animal or person that feeds on seeds.

From Latin *seminis* n., seed + *vorāre*, to eat greedily.

Seeds are among our most enduring, rugged and concentrated foods. They are strong vehicles, designed to carry a plant's offspring to the shore of an uncertain future. If you tear apart a whole grain, bean or nut, you will find within a tiny embryonic shoot. At harvest time, this shoot enters suspended animation (a form of seed hibernation), ready to survive months of drought or cold while waiting for the right moment to come back to life. The bulk of the tissue that surrounds it is a food supply to nourish this rebirth. It is the distillation of the parent plant's life work, a gathering of water, nitrogen and minerals from the soil, carbon from the air, and energy from the sun. As such, it is an invaluable resource for us and other creatures of the animal kingdom that are unable to live solely on soil, sunlight and air. In fact, seeds gave early humans both the nourishment and the inspiration to begin to shape the natural world to their own needs.

Seeds are constructions by which plants create a new generation of their kind. They contain an embryonic plant together with a food supply to fuel its germination and early growth. And they include an outer layer that insulates the embryo from the soil and protects it from physical damage and from attack by microbes or animals. Created to nourish the next generation of living things, they are relatively simple and bland in themselves, but have inspired cooks to transform them into some of the most complex and delicious foods we have.

Most fruit seeds are dispersed when animals eat the fruits and deposit the seeds far from the original source when they defecate. This helps the plants disperse their seeds far and wide, along with a convenient load of fertilizer. Large fruits with large seeds would not have evolved without very large animals to eat them. A big mouth and gut are needed to pass a large, undigested seed. But what happens when the big animals get extinct? The avocado, for example, with its sizeable fruit and pit, would have relied on mammoths and ground sloths, extinct in North and Central America for some 10,000 years now. Generally, when a seed distributor becomes extinct, the plant soon follows. It is hard for a plant species to survive without seed distribution. The avocado had a surprising saviour: jaguars. The felines are large enough to swallow the fruit whole, and the fruit is rich and oily enough to attract their attention as a food source. Dispersion by jaguars kept the avocado viable until its domestication by humans. Another interesting and early example is the *ginkgo*, an ancient tree with an oily fruit that smells like rancid fat. Ginkgoes thrived in the time of the dinosaurs, and some palaeontologists believe that meat-eating dinosaurs were among the animals that distributed gingko seeds.

Listed below some extraordinary seeds and their main qualities:

Poppy seeds come from a west Asian plant, *Papaver somniferum*, which was cultivated by the ancient Sumerians. It is the same plant whose immature seed capsules are cut to collect the latex called opium, a mixture of morphine, heroin, codeine, and other related alkaloid drugs. The seeds are harvested from the capsules after the latex flow has stopped. They may carry traces of opiate alkaloids as well, not enough to affect the body, but enough to cause positive results in drug tests after the consumption of a poppy-flavoured cake or pastry.

Poppy seeds are tiny: it takes 3,300 to make a gram. The seed is 50 per cent oil by weight.

Poppy seeds sometimes have a bitter, peppery taste, the result of damage to the seeds, which mixes oil with enzymes and generates free fatty acids. The striking blue colour of some poppy seeds is apparently an optical illusion. Microscopic examination demonstrates that the actual pigment layer of the seed is brown. Two layers above it, however, is a layer of cells containing

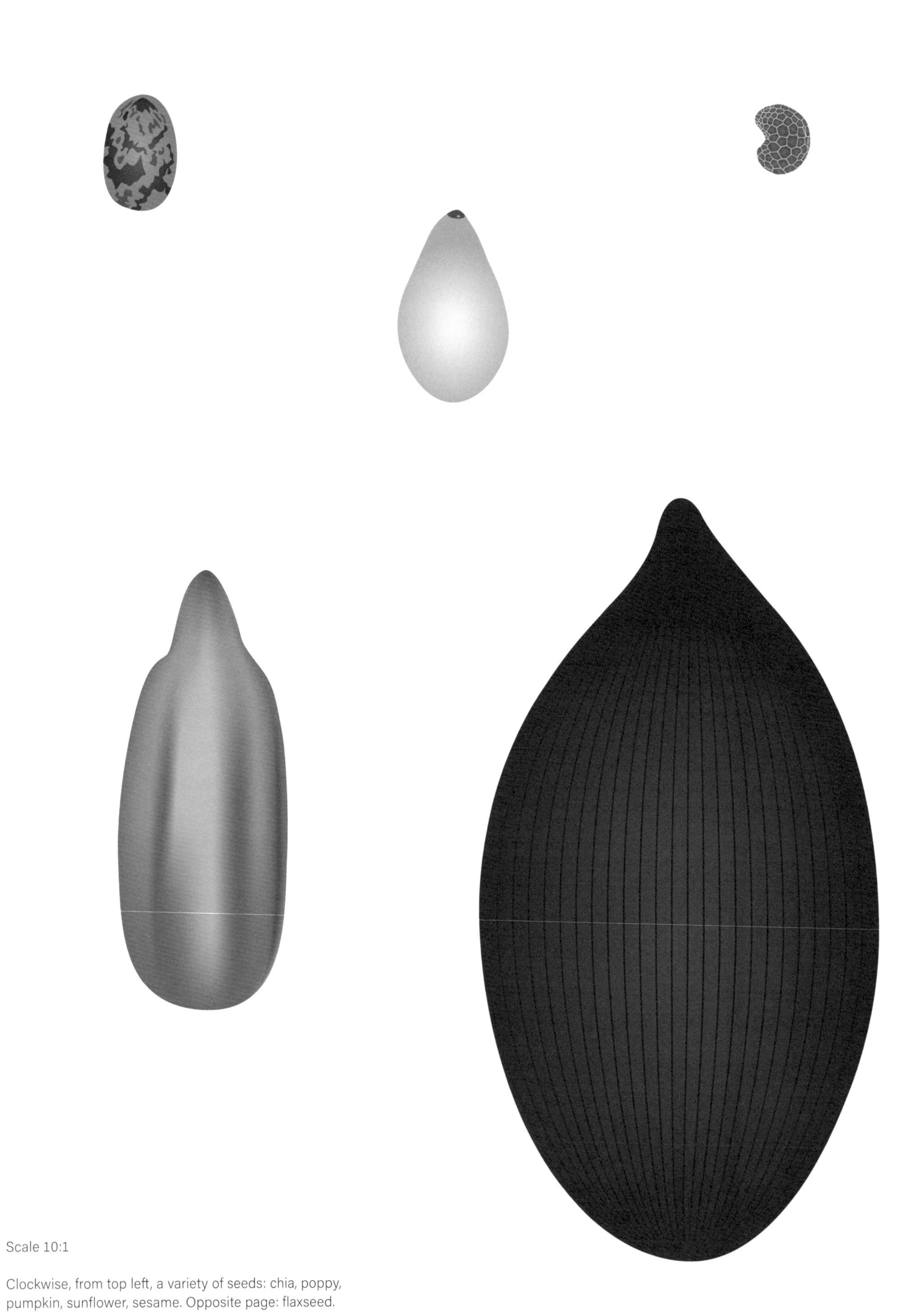

Scale 10:1

Clockwise, from top left, a variety of seeds: chia, poppy, pumpkin, sunflower, sesame. Opposite page: flaxseed.

tiny crystals of calcium oxalate: and the crystals act like tiny prisms, refracting light rays in such a way that blue wavelengths are reflected selectively.

Chia (*Salvia hispanica*) seeds come from a flowering plant originating in southern and central Mexico and Guatemala that is related to the mint family. It is considered to have been as central a crop as maize by economic historians, and thought to have been grown by pre-Columbian Aztecs, as mentioned in the *Codex Mendoza* (16th century). In countries such as Mexico, Argentina, Paraguay, Guatemala and Bolivia, they are consumed, as always, either whole or ground-up in beverages. However, in the last decade chia seeds have been recognized for their beneficial dietary qualities and spread worldwide. An average chia seed is roughly 1 millimetre in diameter and oval in shape, and its colour is a mottled brown, grey, black and white. While soaked, chia seeds can soak up liquid at a rate of up to twelve times their weight and become covered in a jelly-like layer. The seeds give 25–30 per cent extractable oil, including linolenic acid; they are abundant in omega-3 fatty acids.

Pumpkin seeds come from the fruits of the *Cucurbita pepo*, a New World native, and are notable for their deep-green chlorophyll colour, and for containing no starch, as much as 50 per cent oil and 35 per cent protein. Pumpkin seeds are eaten widely as a snack, and in Mexico are used as a sauce thickener. There are so-called naked varieties that lack the usual tough, adherent seed coat and are therefore much easier to cook with. Pumpkin seed oil is a prominent salad oil in central Europe. Intriguingly, different batches of the oil vary in colour. Pumpkin seeds contain both yellowy orange carotenoid pigment and green chlorophyll pigment. Oil pressed from raw seeds is green, but when the seed meal is wetted and heated to increase the yield, more carotenoids are extracted than chlorophyll. The result is an oil that looks dark brown in the bottle or bowl from the combination of orange and green pigments; but in a thin layer, for example on a piece of bread dipped into the oil, where there are fewer pigment molecules to absorb light, the chlorophyll dominates, and the oil becomes emerald-green.

Sunflower seeds come from the flower of the *Helianthus annuus*, the only North American native to become a significant world crop which is a composite of a hundred or more small flowers. Each flower produces a small fruit (like the 'seed' of the strawberry), which is a single seed contained in a thin hull. The sunflower originated in south-western USA, was domesticated in Mexico nearly 3,500 years before the arrival of European explorers, and brought to Europe in around 1510 as a decorative plant. The first large crops were grown in France and Bavaria in the 18th century to produce vegetable oil. Today, the world's leading producer by far is Russia. Improved Russian oil varieties were grown in North America during World War II, and sunflower is now one of the top annual oil crops worldwide. The eating varieties are larger than the oil types, with decoratively striped hulls that are easily removed. Sunflower seeds are especially rich in phenolic antioxidants and vitamin E.

Sesame seeds are the seeds of *Sesamum indicum*, a plant of the central African savanna which is now mostly grown in India, China, Mexico and the Sudan. Sesame seeds are small, with 250–300 per gramme, come in a variety of colours, from golden to brown, violet and black, and are about 50 per cent oil by weight. They are usually toasted lightly to develop a nutty flavour that has some sulphur aromatics in common with roasted coffee. Sesame seeds are made into the seasoned Middle Eastern paste tahini; are added to rice balls and made into a tofu-like cake with arrowroot in Japan; are made into a sweet paste in China; and decorate a variety of baked goods in Europe and the USA. Sesame oil is also extracted from toasted seeds and used as flavouring. The oil is remarkable for its resistance to oxidation and rancidity, which results from high levels of antioxidant phenolic compounds, some vitamin E, and products of the browning reactions that occur during the more thorough toasting.

Flaxseed come from plants native to Eurasia, from the species of *Linum* and especially *L. usitatissimum*, which have been used for more than 7,000 years as a food and to make linen fibre. The small, tough, reddish brown seed is about 35 per cent oil and 30 per cent protein, and has a pleasantly nutty flavour and an attractively glossy appearance. Two qualities set it apart from other edible seeds. First, its oil is over half linolenic acid, an omega-3 fatty acid that the body can convert into the healthful long-chain fatty acids (DHA, EPA) also found in seafood. Flaxseed oil (also known as linseed oil, and valued in manufacturing for drying to a tough water-resistant layer) is by far the richest source of omega-3 fatty acids among plant foods. The seed is about 30 per cent dietary fibre, a quarter of which is a gum in the seed coat made up of long chains of various sugars. Thanks to the gum, ground flaxseed forms a thick gel when mixed with water, is an effective emulsifier and foam stabilizer, and can improve the volume of baked goods.

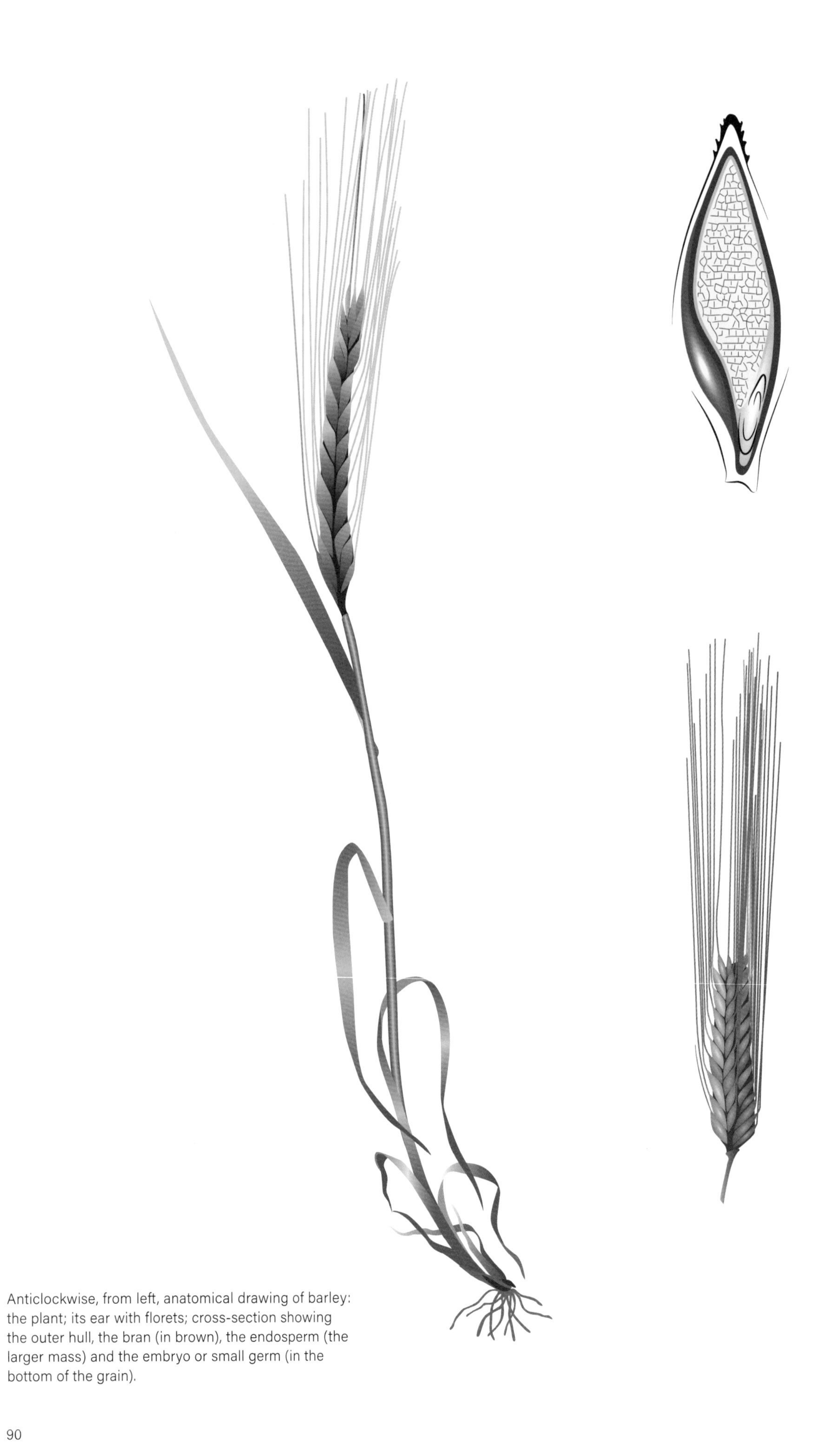

Anticlockwise, from left, anatomical drawing of barley: the plant; its ear with florets; cross-section showing the outer hull, the bran (in brown), the endosperm (the larger mass) and the embryo or small germ (in the bottom of the grain).

Granivore

Granivore *noun*

An animal or person that feeds on grains.

From Latin *granum, grani* n., grain, seed + *vorāre* to eat greedily.

Cereals (from *Ceres*, the Roman goddess of agriculture) or grains are plants in the grass family, and as such thrive in areas that may not be moist enough for trees, like high-altitude grassland or open air. Grains, or *Gramineae*, produce seeds, they are easy to collect and their life cycle is completed in one or two seasons. Grasses are very well suited for agriculture owing to the manner in which they grow, in thick clusters, producing numerous seeds whose great mounts side-line the competition and replace the need for chemical protection. Humans have enabled them to be grown abundantly in large parts of the world, and their use in bread and beer has made them one of our principal forms of nourishment since roughly 3000 BCE. Across the globe different grains have been important in different regions: while in Asia rice has been the most significant, in Africa it has been millets and sorghum, in the New Word maize (corn), and in the Middle East and Europe it has been wheat, oats, barley and rye.

The fact that grains come from the grass family adds a layer of significance to the Old Testament prophet Isaiah's admonishment that 'All flesh is grass'.

Listed below the origin of some of our grains:

Barley (*Hordeum vulgare*) is thought to have been the first domesticated cereal in the south-west Asian grasslands on which wheat grew beside it. Robust and fast growing, one can find it cultivated both in the hot and humid northern Indian plain and in the Arctic Circle. Preceding rice in the Indus valley civilization of western India, it was also the principal cereal in the Mediterranean region and in ancient Egypt, Babylon and Sumer. According to Pliny the Elder, barley was the special food of the gladiators, who were called *hordearii* (barley eaters). Barley porridge, the original polenta, was made with roasted flaxseed and coriander. In the Middle Ages, and especially in northern Europe, barley and rye were the staple foods of the peasantry, while wheat was reserved for the upper classes. Further, *murri*, an Arabian condiment of the Middle Ages similar in flavour to soya sauce (discovered by food historian Charles Perry) was the result of many weeks of fermentation of barley dough.

Today, barley is a minor food in the West; half of production is fed to animals, and a third is used in the form of malt. Elsewhere, barley is made into various staple dishes, including the Tibetan roasted barley flour *tsampa*, often eaten simply moistened with tea. It's an important ingredient in the Japanese fermented soy paste *miso*; and in Morocco (the largest per capita consumer) and other countries of north Africa and western Asia it's used in soups, porridges, and flat breads.

Wheat was one of the first food plants to be cultivated by humans, and was the most important cereal in the ancient Mediterranean civilizations. Wheat was brought to America early in the 17th century, and had reached the Great Plains in the USA by 1855. Compared to other temperate-zone cereals, wheat is a demanding crop. It is susceptible to disease in warm, humid regions, and does best in a cool climate, but it cannot be grown as far north as rye and oats can. A handful of different types of wheat have been grown from prehistoric times to the present. The simplest wheat and one of the first to be cultivated was *einkorn*, which had the standard genetic endowment of most plants and animals: namely two sets of chromosomes. Somewhat less than a million years ago, a chance mating of a wild wheat with a wild goat grass produced a wheat species with four sets of chromosomes, and this tetraploid species gave us the two most important wheats of the ancient Mediterranean world, *emmer* and *durum*. Then, just 8,000 years ago, another unusual mating between a tetraploid wheat species and a goat grass resulted in an offspring with six sets of chromosomes, which in turn gave us our modern bread wheat. The extra chromosomes are thought to contribute to the agricultural and culinary diversity found in

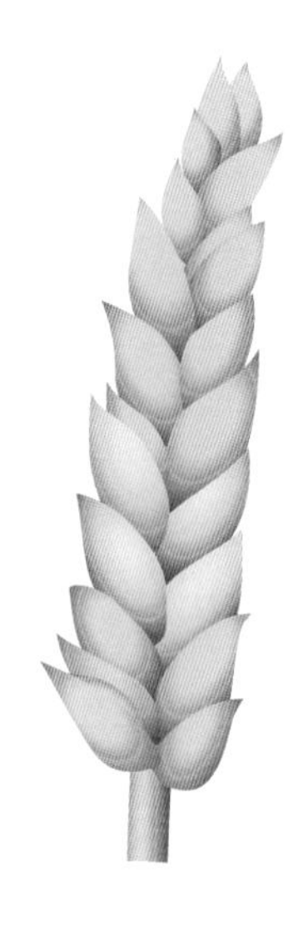

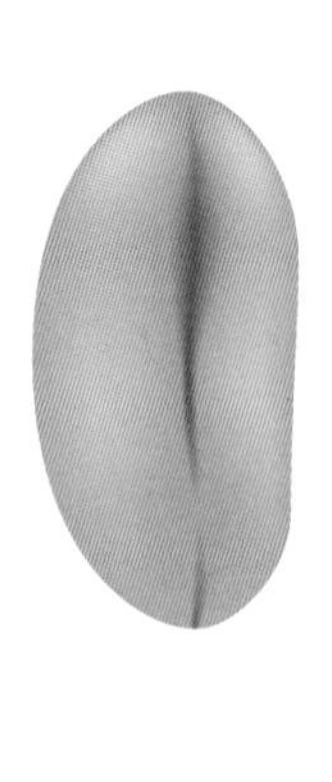

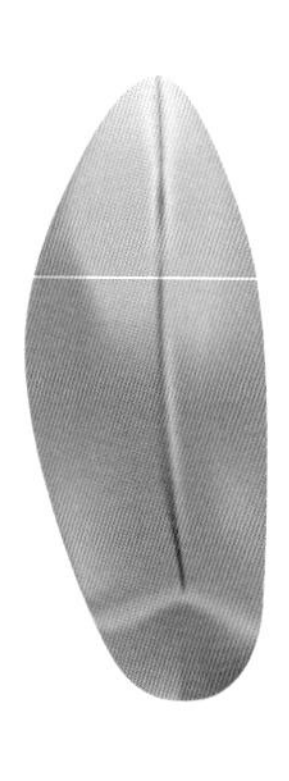

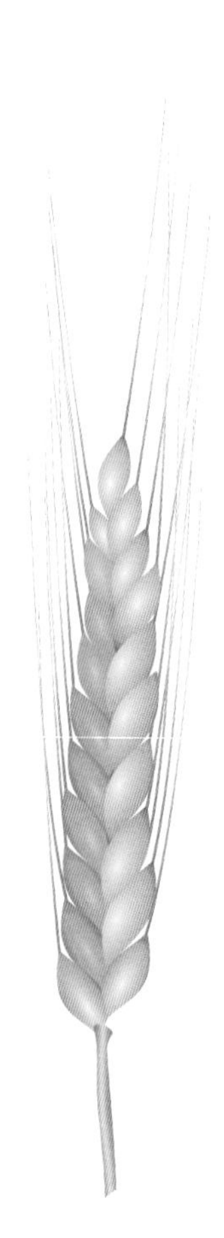

Grain scale 5:1, plant scale 1:1

Clockwise, from top left, a variety of grains: ear and grain of wheat; grain and ear of rye.

modern wheat, most importantly the elasticity of the gluten proteins. Today, 90 per cent of the wheat grown in the world is the hexaploid bread wheat mentioned above. Most of the remaining 10 per cent is durum wheat, whose main purpose is making pasta. The other wheats are still cultivated on a small scale.

Rye (*Secale cereale*) probably comes from south-west Asia, and from there arrived at the Baltic Sea shore at around 2000 BCE (having been carried with barley and wheat by farmers as a weed). Some 1000 years later it was easily cultivated there given that the wet and cool climate and the acid soil suited it more than it did the other cereals. Because it is robust too, it can be found at extremely high altitudes and even in the Arctic Circle. Rye was the most produced bread type in northern Europe until the 20th century. In fact, it was only in 1957 that wheat production exceeded rye in Germany. Today, Poland, Germany and Russia are the biggest producers of rye, and the grain remains especially popular in Finland, Scandinavia and Eastern Europe.

The unusual carbohydrates and proteins in rye are the reason it produces a distinctive kind of bread. Rye contains a large quantity – up to 7 per cent of its weight – of carbohydrates called pentosans, thanks to which rye flour absorbs 8 times its weight in water, while wheat flour absorbs 2. Unlike starch, pentosans do not retrograde and harden after being cooked and cooled, and so they provide a soft, moist texture that helps give rye breads a shelf life of weeks.

Maize (*Zea mays*) originates from teosinte (Zea mexicana) an open woodland large grass that was domesticated in Mexico some 7,000–10,000 years ago. In the USA it is called corn. While some legumes and cereals of the Old World were changed relatively little, the very form of teosinte was altered in extreme manners. The large size and juicy fruit that came from the cross- breeding made, corn to be the essential food plant of many of the natives of early America: the Mayas and Aztecs of Mexico, the Incas of Peru, the cliff dwellers of south-western USA, Mississippi mound builders, and numerous South and North American semi-nomadic people.

Corn began being grown in southern Europe within a generation after Christopher Columbus brought the crop back. Now, It is the basic food for millions in Africa, Asia and Latin America, as well as being the third largest food crop for humans (after wheat and rice). In the USA and Europe more corn is consumed by livestock than by humans, but still people enjoy it as a snack or cooked in different ways. Additionally, it is made into oil, used as a filling and to thicken sauces because of its starch content, as mash in whiskey production and some sweet preparations. Corn steep liquor and ethanol are some of the industrial items made from other parts of the corn plant.

Rice is the principal food for about half the world's population, and in such countries as Bangladesh and Cambodia provides nearly three-quarters of the daily energy intake. *Oryza sativa* is a native of the tropical and semi-tropical Indian subcontinent, northern Indo-China and southern China, and was probably domesticated in several places independently: the short-grain types around 7000 BCE in the Yangtze river valley of south-central China, and the long-grain types in South East Asia somewhat later. A sister species with a distinctive flavour and red bran, *Oryza glaberrima*, has been grown in West Africa for at least 1,500 years.

The grain found its way from Asia to Europe via Persia, where the Arabs learned to grow and cook it. The Moors first grew large quantities of rice in Spain in the eighth century, then somewhat later in Sicily. The valley of the Po river and the Lombardy plain in northern Italy, the home of *risotto*, first produced rice in the 15th century. The Spanish and Portuguese introduced rice throughout the Americas in the 16th and 17th centuries. South Carolina was the location of the first commercial planting in the USA, in 1685, where the rice-growing expertise of African slaves was important.

There are thought to be more than 100,000 distinct varieties of rice throughout the world. The majority of rice is milled to remove the bran and most of the germ, and then polished with fine wire brushes to grind away the aleurone layer and its oil and enzymes. The result is a very stable refined grain that keeps well for months.

Millet is the name used for a number of different grains, all containing small round seeds, 1–2 millimetres in diameter (the species of *Panicum, Setaria, Pennisetum* and *Eleusine*). The millets are native to Africa and Asia, and have been cultivated for 6,000 years. They are especially important in arid lands because they have one of the lowest water requirements of any cereal, and will grow in poor soils. The grains are remarkable for their high protein content, from 16 to 22 per cent, and are popped like corn but also made into porridge, breads, malts and beers.

Oats are produced in greater quantities than rye worldwide today, but 95 per cent of the crop is fed to animals. They are the grain of *Avena sativa*, a grass thought to have originated in south-west Asia and gradually come under cultivation as a companion of wheat and barley. In Greek and Roman times, it was considered a weed or a diseased form of wheat. By 1600, it had become an important crop in northern Europe, in whose wet climate it does best, for oats require more moisture than any other cereal but rice. Some ambivalence towards the grain may have lingered, however, as evidenced by Samuel Johnson's dictionary, published in 1755, which defines oats as: 'A grain, which in England is generally given to horses, but in Scotland supports the people.'

Grain scale 5:1, plant scale 1:1

From left: ear and grain of maize, commonly known as corn.

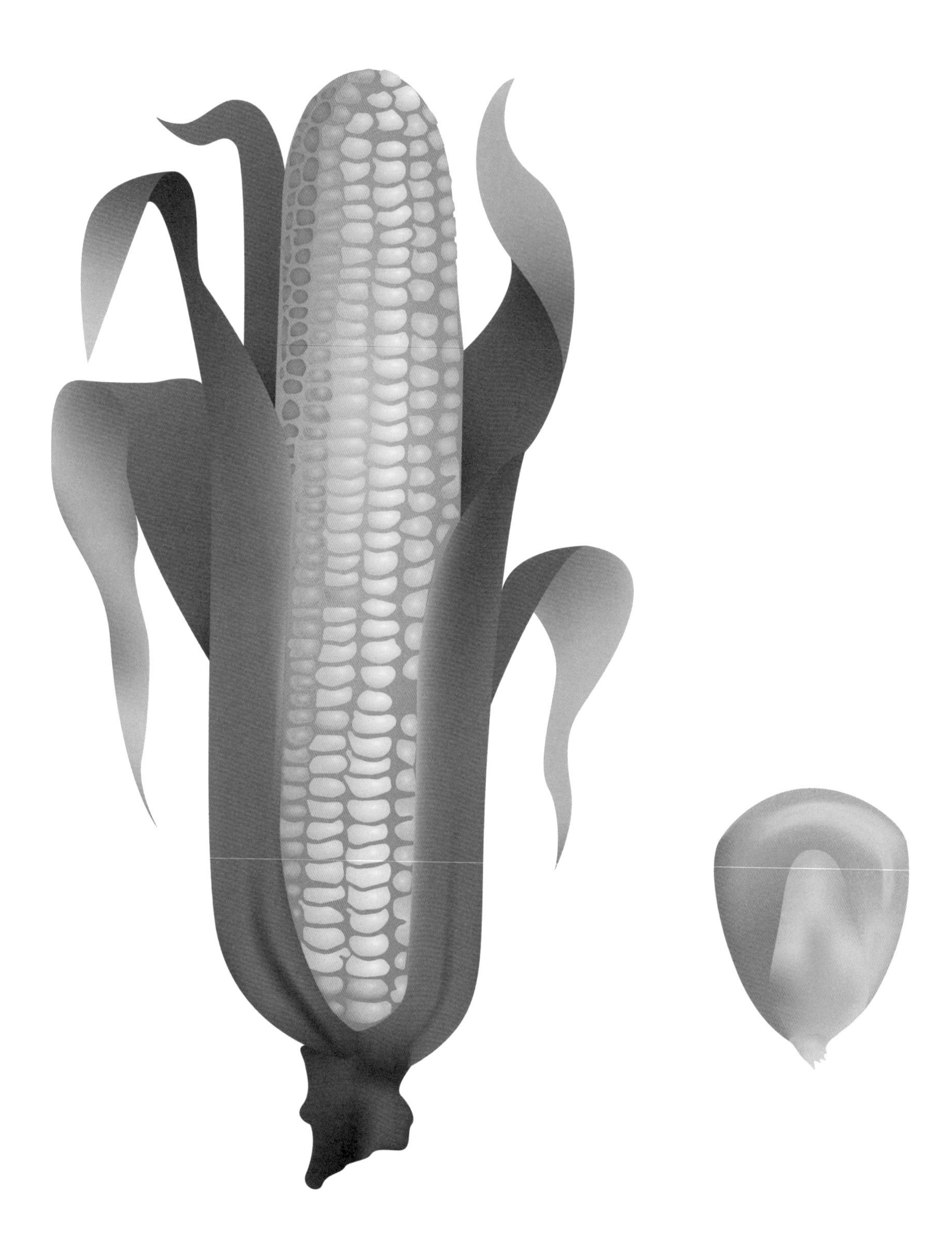

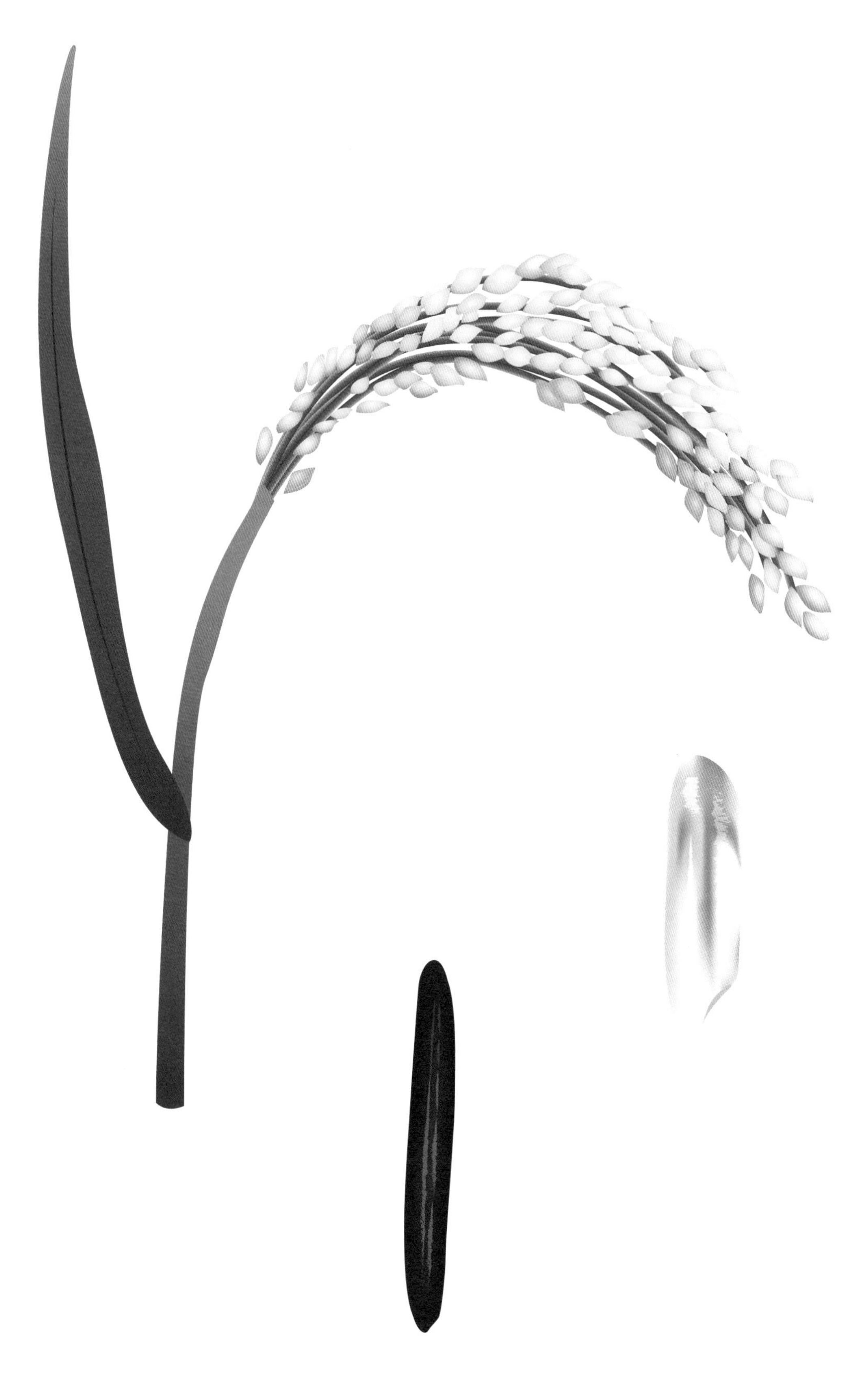

Grain scale 5:1, plant scale 1:1

From left: rice plant, a wild rice kernel and a polished rice kernel.

Today, the UK and the USA are the largest consumers of food oats. Consumption in the USA was boosted in the late 19th century by Ferdinand Schumacher, a German immigrant who developed quick-cooking rolled oats for breakfast, and Henry Crowell, who was the first to turn a cereal from a commodity into a retail brand by packaging oats neatly with cooking instructions, labelling them 'Pure', and naming the product Quaker Oats. Oats are now a mainstay in ready-to-eat granola, muesli and manufactured breakfast cereals. They are rich in indigestible carbohydrates called beta-glucans, which absorb and hold water, give hot oatmeal its smooth, thick consistency, have a tenderizing, moistening effect in baked goods, and help lower our blood cholesterol levels. Oats also contain a number of phenolic compounds that have antioxidant activity.

Buckwheat (*Fagopyrum esculentum*) is a plant in the *Polygonum* family, a relative of rhubarb and sorrel. It is a native of central Asia, was domesticated in China or India relatively recently, around 1,000 years ago, and was brought to northern Europe during the Middle Ages. It tolerates poor growing conditions and matures in a little over two months, so it has long been valued in cold regions with short growing seasons.

Buckwheat kernels are triangular, 4–9 millimetres across, with a dark hull. The inner seed is a mass of starchy endosperm surrounding a small embryo, and is contained in a light greenish yellow seed coat. Intact seeds with the hull removed are called groats. Buckwheat is about 80 per cent starch and 14 per cent protein. It contains approximately double the oil of most cereals, which limits the shelf life of groats and flour. The distinctive aroma of cooked buckwheat has nutty, smoky, grassy and slightly fishy notes.

Quinoa was a basic Inca form of sustenance (nearly as important as potatoes). It originated in northern South America and cultivated near Lake Titicaca in the Andes some 7,000 years ago. The diameter of these spherical yellow grains is between 1 and 3 millimetres, and the outer pericarp of numerous varieties contains bitter saponins, protective compounds that may be rubbed and washed off with cold water. *Chenopodium quinoa* is related to spinach and beets. Quinoa in general is usually added to soups or cooked like rice, as well as being popped or made into flour for flatbread.

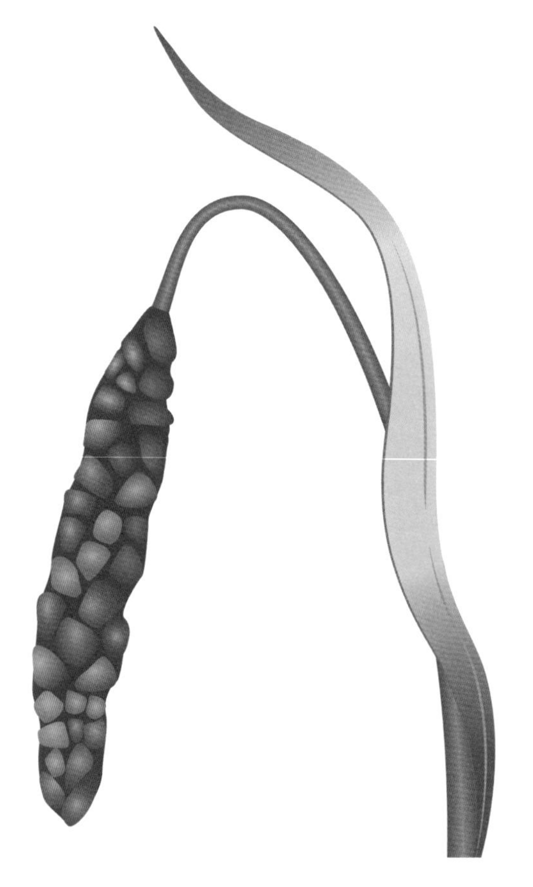

Grain scale 5:1, plant scale 1:1

Opposite page, from left: millet grain and plant. Above, left to right, from top: oat ear and grain; buckwheat grain and ear.

From left: a cluster of varicoloured pistachios in their hulls on the tree, and a pistachio kernel in its shell.

Nucivore

Nucivore *noun*

An animal or person that feeds on nuts.

From Latin *nux, nucis* f., nut + *vorāre*, to eat greedily.

Nuts originate from a variety of plant families and the name comes from the Indo-European word for 'compressed'. They grow on long-lived trees and usually comprise a seed encircled by a tough shell. Their relatively large size has a dual function: to allow enough sustenance during the long growth process and to appeal to animal dispersers who may forget about some buried items. Most nuts store their energy not in starch but in oil, which is a more dense chemical form.

Nuts are much less important in the human diet than grains or legumes because their trees do not begin to bear nuts until years after they are planted, and cannot produce as much per hectare as the quick-growing grains and legumes. The biggest exception to this rule is the coconut, a staple food in many tropical countries. Another is the peanut, which is a legume with an uncharacteristically oily, tender seed, and which can be grown quickly in massive numbers.

Listed below are a selection of important nuts:

Pistachios are the seeds of a tree, native to western Asia and the Middle East; *Pistacia vera*, a relative of the cashew and the mango. Along with almonds, they have been found at the sites of Middle Eastern settlements dating to 7000 BCE. A close relative, *Pistacia lentiscus*, provides the aromatic gum called *mastic*.

> *Pistachios first became a prominent nut in the USA in the 1880s, thanks to their popularity among immigrants in New York City.*

Iran, Turkey and California are the major producers today. Pistachios grow in clusters, with a thin, tannin-rich hull around the inner shell and kernel. As the seeds mature, the outer hull turns purplish red and the expanding kernel cracks the inner shell open. Traditionally, the ripe fruits were knocked from the trees and sun-dried, and the hull pigments stained the shell, so the shells were often dyed to make them a uniform red. Today, most California pistachios are hulled before drying, so the shells are their natural pale tan colour.

Pistachios are remarkable among the nuts for having green *cotyledons* (a cotyledon is an embryonic leaf in seed-bearing plants). The colour comes from chlorophyll, which remains vivid when the trees grow in a relatively cool climate, for example at high elevation, and when the nuts are harvested early, several weeks before full maturity. Pistachios thus offer not only flavour and texture but also a contrasting colour in pâtés, sausages, other meat dishes and in ice creams and sweets. The colour is best retained by roasting or otherwise cooking the kernels at low temperatures that minimize chlorophyll damage.

Walnuts from the genus *Juglans* come from around the fifteen species native to south-western Asia, East Asia and the Americas. *J. regia*, or the Persian or English walnut, is domesticated in the greatest quantities, its seeds having been consumed since antiquity in Europe and western Asia. Apart from almonds, walnuts are the most widely eaten tree-nuts globally; indeed, the words for nut and walnut are the same in several languages of Europe. The USA, France and Italy are the major producers today. The culinary uses for walnuts vary from oil and, in China and Europe, milk, to providing the basis for sauces such as *satsivi* in Georgia, *fesenjan* in Persia and *nogada* in Mexico. Young walnuts flavour sweetened alcohol such as Sicilian *nocino* or French *vin de noix*, are yielded in early summer and pickled in Britain, or preserved in syrup the Middle East.

Almonds are the world's largest tree-nut crop. They are the seeds of a plum-like stone fruit; in fact, the tree is a close relative of the plum and peach. There are several dozen wild or minor species, but the cultivated almond, *Prunus amygdalus*, came from western Asia and had been domesticated in the Bronze Age. California is now the largest producer. Thanks to their high content of antioxidant vitamin E and low levels of polyunsaturated fats, almonds have a relatively long shelf life.

Pine nuts are gathered from about a dozen of the one hundred species of pines, one of the most familiar evergreen tree families in the northern hemisphere. Among the more important sources are the Italian stone pine *Pinus pinea*, the Korean or Chinese pine *P. koraiensis* and the south-western USA pines *P. monophylla* and *P. edulis*.

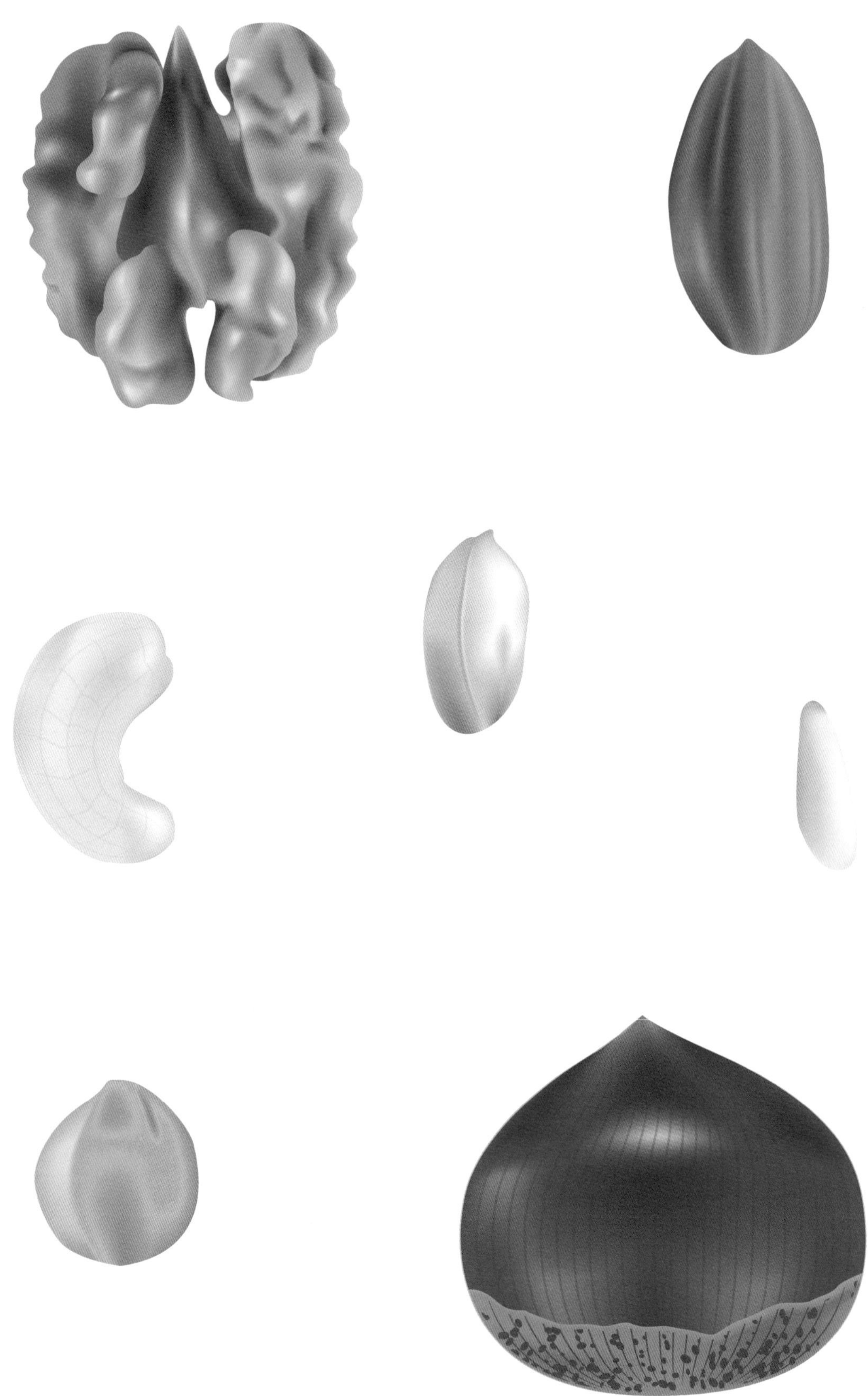

Scale 2:1

Clockwise, from top left, a variety of nuts: walnut, almond, pine nut, chestnut, hazelnut, cashew, peanut. Opposite page: brazil nut.

The nuts are born on the scales of the pine cone, which takes three years to mature. The cones are sun-dried, threshed to shake out the seeds, and the kernels then hulled, nowadays by machine. They have a distinctive, resinous aroma and are rich even for nuts; Asian pine nuts have a higher oil content (78 per cent) than either USA or European types (62 per cent and 45 per cent, respectively). They are used in many savoury and sweet preparations, and pressed to make oil. In Korea, pine pollen is used to make sweets, and Romanians flavour game sauces with the green cones.

Chestnuts from the large tree *genus Castanea* are found in Europe, Asia and North America. The energy stored in them for seedlings ahead takes the form of starch, rather than the more typical oil in other nuts. Their starch content also accounts for their crumbly composition and why they are mostly eaten cooked. Further, their varied applications go back millennia: as soup thickeners, dried, made into flour for cakes, breads, gruels and pastas. While being a staple in the rural and mountainous regions of Switzerland, France and Italy before being superseded by the New World corn and potato, they are conversely also made into a luxurious sweet. Invented in the 17th century, *marrons glacés* are large cooked chestnuts that are infused with a vanilla-flavoured syrup over some forty-eight hours and finally glazed with a condensed syrup.

Hazelnuts from the *genus Corylus* come from a few of the fifteen species of predominantly bushy northern-hemisphere trees. In prehistoric times, the Eurasian natives *Corylus avellana* and *C. maxima* were utilized for their nuts as well as for the making of walking sticks from their shoots. Now, around the Turkish Black Sea region, the much taller *C. colurna* is responsible for most of these nuts. In the UK, the longer-shaped variety is called a 'filbert', perhaps a reference to St. Philibert's Day in late August, around which time hazelnuts begin to mature. In cooking hazelnuts have many uses: Apicius's cookbook included them in sauces for dishes containing mullet, birds and boar; in Spain they can be used in picada and romanesco sauces instead of almonds; in Italy they are used in a liqueur called *Frangelico*; while in Egypt they are included in the condiment *dukka*.

Cashews are nuts that come from the cashew tree (*Anacardium occidentale*) Amazon region, like brazil nuts, and are some of the most popular nuts globally, with the biggest modern-day producers being East Africa and India, to which they were brought by the Portuguese. The oil in its shell contains an irritant oil (the nut is related to poison ivy) that must be expelled by heating. Interestingly, the seed-containing fruit shell tends to be removed and discarded in producing countries while the enlarge stem tip or 'false fruit', called the cashew apple, is consumed in a variety of ways: fresh, cooked or fermented into an alcoholic drink.

Peanuts are the seeds of *Arachis hypogaea*, a small bush, and its fruit pods are pushed underground as the plant grows. It is related to the bean family and is, in fact, not a nut at all. Some 4,000 years ago it was cultivated in South America, most likely Brazil, and even prior to the Incas was a central food crop for Peruvians. Once carried to India, Asia and Africa by the Portuguese in the 16th century, it became an important cooking oil in China (peanuts have double the oil content of soy beans).In USA peanuts where used to feed animals for centuries, it was only in the early 1900s that peanuts began to supplant weevil-ravaged cotton, on the advice of agricultural scientist George Washington Carver.

Brazil nuts are unusually large, 2.5 centimetres or more long, and double the weight of almonds and cashews. They are the seeds of a large tree (*Bertholletia excelsa*, 50 metres tall and 2 metres across) native to the Amazon region of South America, and develop in clusters of eight to twenty-four inside a hard, coconut-sized shell. South American countries are still the main producers. The pods are gathered only after they fall to the ground. Because they weigh about 2 kilogrammes each, the pods can be lethal missiles, and harvesters must carry shields to protect themselves. The edible portion of the seed is an big swollen embryonic stem. Thanks to their size and high oil content, two large Brazil nuts are the caloric equivalent of one egg.

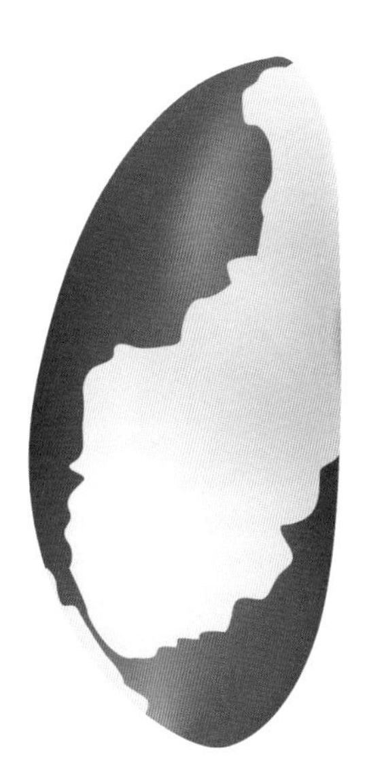

The lentil plant has beautiful bright-green leaves and the typical curly tendrils that help the plant to climb to sunny heights by holding onto objects in their surroundings.

Leguminivore

Leguminivore *noun*

An animal or person that feeds on legumes.

From Latin *legumen, leguminis* n., legume + *vorāre*, to eat greedily.

Legumes, like other plants in the bean family, the *Leguminosae*, bear pods that contain several seeds. Their name comes from the Latin *legere*, 'to pick', and the term legume is also used to name their seeds. The majority of varieties are vines that seek the maximize sunlight by climbing on adjacent plants; they also take only several months to complete their life cycle. Their symbiosis with their root bacteria that feed them with nitrogen from the air not only means that their seeds are highly rich in protein but also that they enrich with nitrogen compounds the soil in which they grow, allowing legumes to be grown as rotation crops since antiquity. Their pods are covered in camouflage patterns, and have a variety of biochemical protection available to them. Further, they appeal to different animals because of their size, and the need to protect themselves from insects is considered to be the reason for the exceptional variety in peas and beans.

Of the flowering plants, legumes belong to the third largest family (after the orchid and daisy), and to the second most important family in the human diet, after the grasses (which include grains). Peas and beans have long acted as high protein replacements for expensive animal foods, particularly in the Mediterranean, Asia and Central and South America. The naming of the four most important Roman families is indicative of their place in antiquity: *Fabius* comes from the fava bean, *Lentulus* from the lentil, *Piso* from the pea and *Cicero* – most distiguished of them all – from the chickpea. Today, there are some twenty legume species that are mass grown. However, it is the types of uses to which a legume is put that determine its level of production: next to being food for us, peanuts and soya beans are also used as animal feed and for oil (both kitchen and industrial) and are therefore more widely grown than legumes that are only as food for humans.

Listed below are some of the large variety of beans available:

Lentils are probably the oldest cultivated legume, contemporaneous with wheat and barley, and often growing alongside these grasses. Their native ground is around south-west Asia, and they are now commonly eaten across Europe and Asia. Nowadays, most lentils are produced in India and Turkey, with Canada a distant third. The Latin word for lentil, *lens*, gives us our word for a lentil-shaped or doubly convex piece of glass (the coinage dates from the 17th century). Lentils contain low levels of antinutritional factors (which interfere with the absorption of diet-based minerals) and cook quickly.

Kidney beans are a variety of the common bean. They are named for their resemblance in shape and colour to kidneys, have a soft and creamy flesh and are available dried or tinned. Dried kidney beans need soaking and should be cooked carefully because they contain toxins on the outer skin when raw, which are rendered harmless by boiling. They are great in mixed bean salads and stews such as chilli con carne.

Fava beans or broad beans, *Vicia faba*, are the largest of the commonly eaten legumes, and were the only beans known to Europe until the discovery of the New World. They apparently originated in west or central Asia, and were among the earliest domesticated plants. Larger cultivated forms have been found in Mediterranean sites dating to 3000 BCE. There are several sizes of broad bean, the largest of which seems to have been developed in the Mediterranean region around 500 CE. China is the world's largest producer. Broad beans can be eaten raw, when double hulled, or cooked, and have a fresh grassy note to them. When double hulled the fava bean has a bright green colour.

Lima beans (named after the Peruvian capital) preceded the consumption of the common bean in Peru, which was native to Central America. Lima beans are larger than common beans. Both species arrived in Europe with Spanish explorers. Having arrived in Africa via the slave trade, they are now the main legume of its tropics. While certain tropical varieties and the wild kind require cooking to remove the possibly toxic levels of a cyanide defence system, shop bought varieties do not contain cyanide and the cultivated beans can be consumed either dried or fresh.

Azuki or adzuki (*Chinese chi dou*) is an East Asian species of *Vigna*, called *V. angularis*, and most commonly has a deep maroon colour, which makes it a favourite ingredient for festive occasions, for in Asian cultures such as China and Japan red is associated with good fortune. It was cultivated in Korea and China at least 3,000 years ago, and taken later to Japan; it is now the second most important legume after the soya bean in both Korea and Japan. Azuki are a favourite sprouting seed, and in Asia are also candied, infused with sugar to make a dessert topping and used as a base for a hot drink. In Japan, most of the azuki crop is made into a sweet paste composed of equal parts sugar and ground twice-boiled azuki, which are kneaded together.

Tepary beans, small brown natives of south-west USA, are unusually tolerant of heat and water stress. They are especially rich in protein, iron, calcium and fibre, and have a distinctive, sweet flavour reminiscent of maple sugar or molasses.

Cranberry beans, also known as *borlotti* in Italy and shell beans in some other regions, are high in nutritional value and make a great addition to a number of dishes. They get their name from the appearance of their pods, which are often red or pink. The beans themselves are usually white or cream in colour with deep red specks, which typically disappear as they darken during cooking. Despite their name, cranberry beans are not related to cranberries and resemble pinto beans in terms of texture and size.

Black-eyed beans, or cowpeas in the USA, are an African relative of the mung bean that were known to ancient Greece and Rome and brought to the southern USA with the slave trade. They have an eye-like anthocyanin pigmentation around the hilum, and a distinctive aroma. A variety that produces an elongated pod and small seeds is the yard-long bean, a common green vegetable in China.

Soya beans are produced on a larger scale than most legumes worldwide, and are used for numerous purposes, from animal feed to tofu production. They are also eaten before they are fully mature, and are therefore sweeter, contain lower levels of gassy and antinutritional substances and have a less pronounced beany flavour. Fresh soya beans, Japanese *edamame* or Chinese *mao dou*, are specialized varieties harvested at 80 per cent maturity, still sweet and crisp and green, then boiled for a few minutes in salted water. Green soya beans contain approximately 15 per cent protein and 10 per cent oil.

Grams, from the genus *Vigna*, which is native to the Old World, provide the small-seeded grams of India and a few other Asian and African seeds. Most of them have the advantages of being small, quick cooking and relatively free of anti-nutritional and discomforting compounds. Green grams, or mung beans, are native to India, and spread early on to China. Thanks to the popularity of their sprouts, they are now the most widely grown of this group. Black gram, or *urad dal*, is the most prized of the legumes in India, where it has been cultivated for more than 5,000 years and is eaten either whole, or split and de-hulled, or ground into flour for cakes and breads.

Peas have been cultivated for around 9,000 years and spread quite early from the Middle East to the Mediterranean, India and China. They are a cool-climate legume that grows during the wet Mediterranean winter and in the spring of temperate countries. They were an important source of protein in Europe in the Middle Ages and later, as the old nursery rhyme attests: 'Pease porridge hot / Pease porridge cold / Pease porridge in the pot / Nine days old.' Today, two main varieties are cultivated: a starchy, smooth-coated one that gives us dried and split peas, and a wrinkly type with a higher sugar content, which is usually eaten when immature as a green vegetable. Peas are unusual among legumes in retaining some green chlorophyll in their dry cotyledons; their characteristic flavour comes from a compound related to the aroma compound in green peppers, and their nutritional values are high. Like fava beans, peas can be eaten raw and fresh as well as boiled.

Chickpeas or *Garbanzos* have been grown since the 7th century BCE, similarly to lentils, fava beans and peas, and originate in dry south-west Asia. There are two types: *desi*, which is dark in colour, small, has a thick seed coat and is close to the wild variety; and *kabuli*, which is cream-coloured, is larger, and has a thin, light seed coat. The former is the main variety grown in Mexico, Asia, Ethiopia and Iran, while the latter is typical in the Mediterranean and Middle East. Whilst the majority of legumes are 1–2 per cent oil by weight, chickpeas are nearer to 5 per cent, and their nutty and creamy flavour makes them ideal for dips and salads.

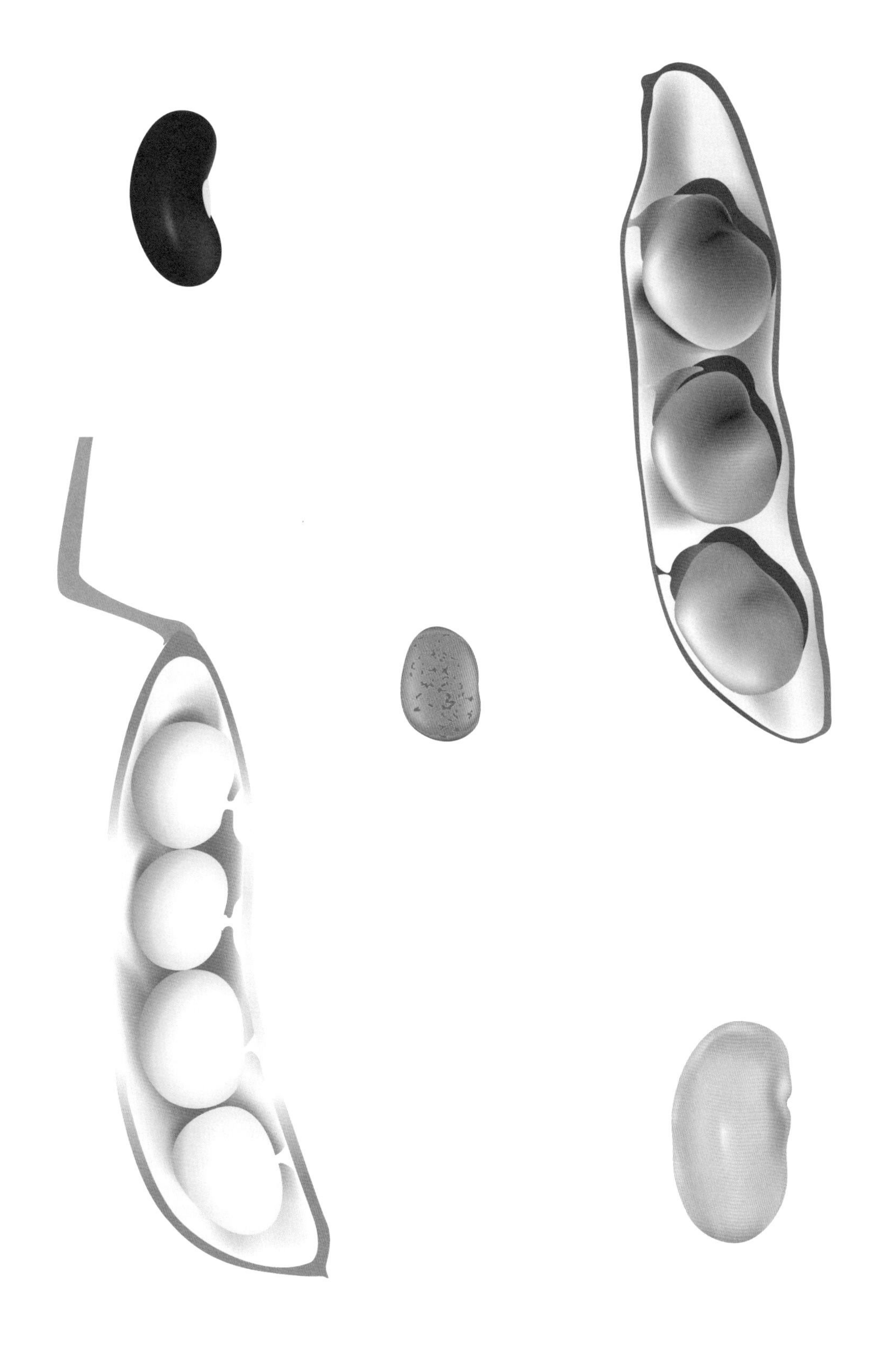

Scale 1:1

Clockwise, from top left, a variety of legumes: kidney, fava, butter, lima, pinto.

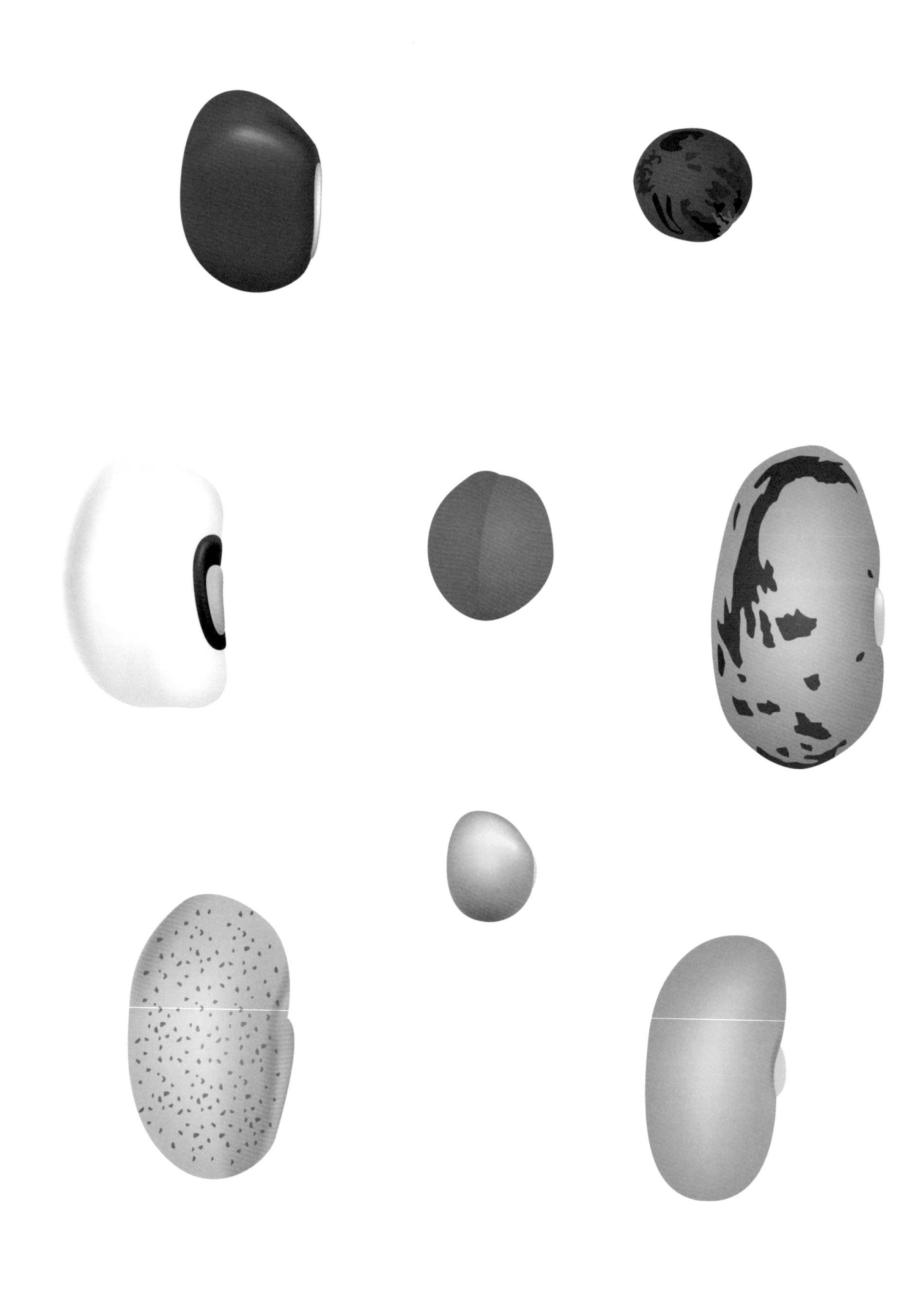

Scale 3:1

Clockwise, from top left, a variety of legumes: azuki, green lentil, cranberry, white bean, white cranberry bean, black eyed pea, soya, green gram.

The common bean (*P. vulgaris*) is the most important species of the genus *Phaseolus*, which originated in south-western Mexico, where it continues to be eaten more widely than in other parts of Latin America. Domestication began at around the 5th century BCE, moving slowly both north and south, and it has grown into many hundreds of varieties that differ in size, shape, seed-coat colour and colour pattern, shininess and flavour. Types containing larger seeds (kidney, cranberry, large red and large white) derived from the Andes, and became established in north-east USA, and in Europe and Africa, while varieties containing smaller seeds from Central American (pinto, black, small red and small white) were concentrated in south-west USA. There are more than a dozen commercial categories based on colour and size in the USA, and they are cooked in many ways, varying from pastes and desserts to boiled and made into soups and stews.

Domesticated in the high Andes for several thousand years, the *nuña*, also known as popping bean, can indeed be popped like pop-corn, for 3–4 minutes on a high dry heat, which is important in the fuel-poor mountains. Unlike popcorn, its consistency is floury and it remains compact and expands only a little.

Pigeon peas are distant relatives of the common bean; they are native to India and are now grown throughout the tropics. In India, they are called *toor dal* or red gram because the tough seed coat of many varieties is reddish brown, though it is most often hulled and split, and the cotyledons are yellow or light green. They have been cultivated for around 2,000 years, and are made into a simple porridge. Like the other grams, they contain little antinutritional factors.

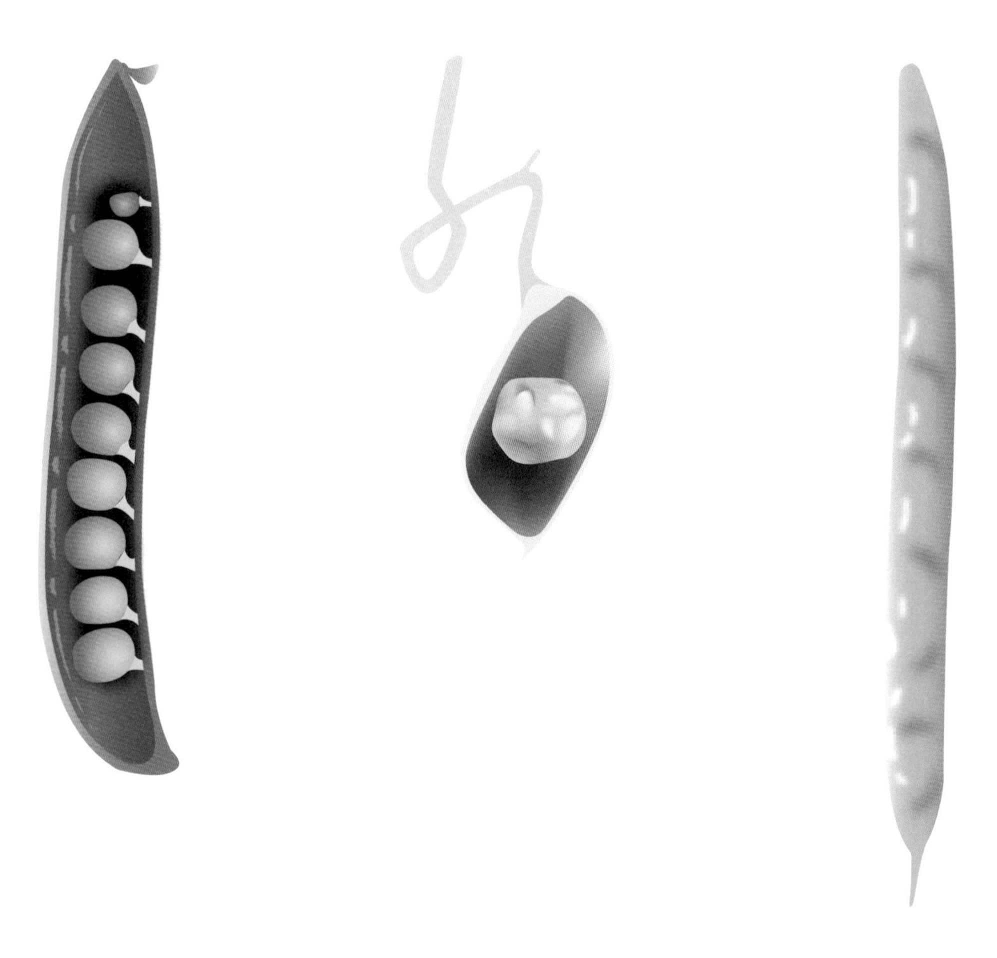

Scale 1:1

From left, a variety of legumes: green peas, chickpea, common bean.

Result

The Art of Mixing & Sausage Matrix

So far, we have explored various aspects of sausage making: the origins of sausages, construction processes and possible future ingredients. Yet a final element remains, one that is essential to the creation of a future sausage: the mixture.

117 Futurivore
147 The Psychology of Disgust and the Desire for Delicacy

At the Ter Weele butchery, Netherlands: during a test for a mortadella with vegetables. The vegetables were first blanched, then dusted with flour to absorb moisture.

The Art of Mixing

Having looked thoroughly at the construction processes of the sausage, it is now time to consider one final essential element in the creation of a future sausage: the mixture. To meet all the requirements of a sausage, its mass must consist of a combination of ingredients in a certain ratio. When meat is minced, mixed with salt and packed together, the leaking myosin binds the meat together upon cooking. Without this mincing or mixing with salt, the myosin cannot do its job. Furthermore, when making a fresh sausage, it is important for it to contain at least 30 per cent moisture, often in the shape of fat, so that it will keep the other lean meat moist and tender during the cooking process. Apart from these functional aspects of mixing, there is one far more complex and intriguing element of the mixture: flavour. Of all meat preparations, the sausage is the one that has the most complex and diverse array of flavours and the most delectable textures, and is rarely dry or under-seasoned.

Our flavour experience is predominantly made up of the principal sensations of taste, touch and smell; sight also plays a key role, for example through the impact of the colour or the presentation of our food. An expectation of crunchiness can affect our perception and even our emotions; the memory of a bitter chocolate can be intwined with that of a broken heart; and the taste of savoury roasted sausages can instantly transport us back to a picnic in the mountains.

> *The way we register the flavour of our food is composed of several distinct sensations.*

With the taste buds on our tongues, we register salts, sweet sugars, sour acids, savoury amino acids and bitter alkaloids. On the other hand, the cells in our mouth are sensitive to touch, which is why we may notice the presence of the sharp tannins. Also, the pungent compounds in peppers, mustard, and members of the onion family can irritate a variety of cells in and near the mouth. Finally, the olfactory receptors in our nasal passages can detect many hundreds of volatile molecules. The sensations from our mouth give us an idea of a food's basic composition and qualities, while our sense of smell allows us to make much finer distinctions.

Nearly all the aromas present in our food are composed of many different volatile molecules. In the case of vegetables, herbs and spices, the number may be a dozen or two, while fruits typically emit several hundred volatile molecules – although usually it is just a handful which create the dominant element of an aroma. For these finer distinctions, there exists a slightly more subtle vocabulary with terms such as fruity, nutty, grassy, citrus, earthy, meaty, woody, spicy, and so on. Sometimes it is difficult to question the rapid assumptions we make about our food; surely steak just tastes like steak, right? In fact, the aromatic qualities of ingredients are far more surprising and unexpected. For example, peas are fruity, pork liver is roasted and herbal, raw scallops are floral and steak actually has a very dominant cheesy note. These aromas also reveal which ingredients fit well together and which clash. For example, strawberry has an apparent cheesy note that makes it match well with cheeses such as Parmesan, whereas its citrusy note matches basil perfectly. Sometimes even stranger seeming ingredients fit perfectly together, such as blue cheese and dark chocolate or scallops and anise, whilst the molecular balance of others is questionable, such as smoked salmon and strawberry or basil and coffee. The latter combination increases the bitterness of coffee to a level unappealing to most.

The art of making a future sausage is to know your ingredients and methods; it is to construct a combination of ingredients that tick all the boxes, by trial and error. It is essential to have an open view on what is considered food and leave behind assumptions of flavour to really begin tasting again.

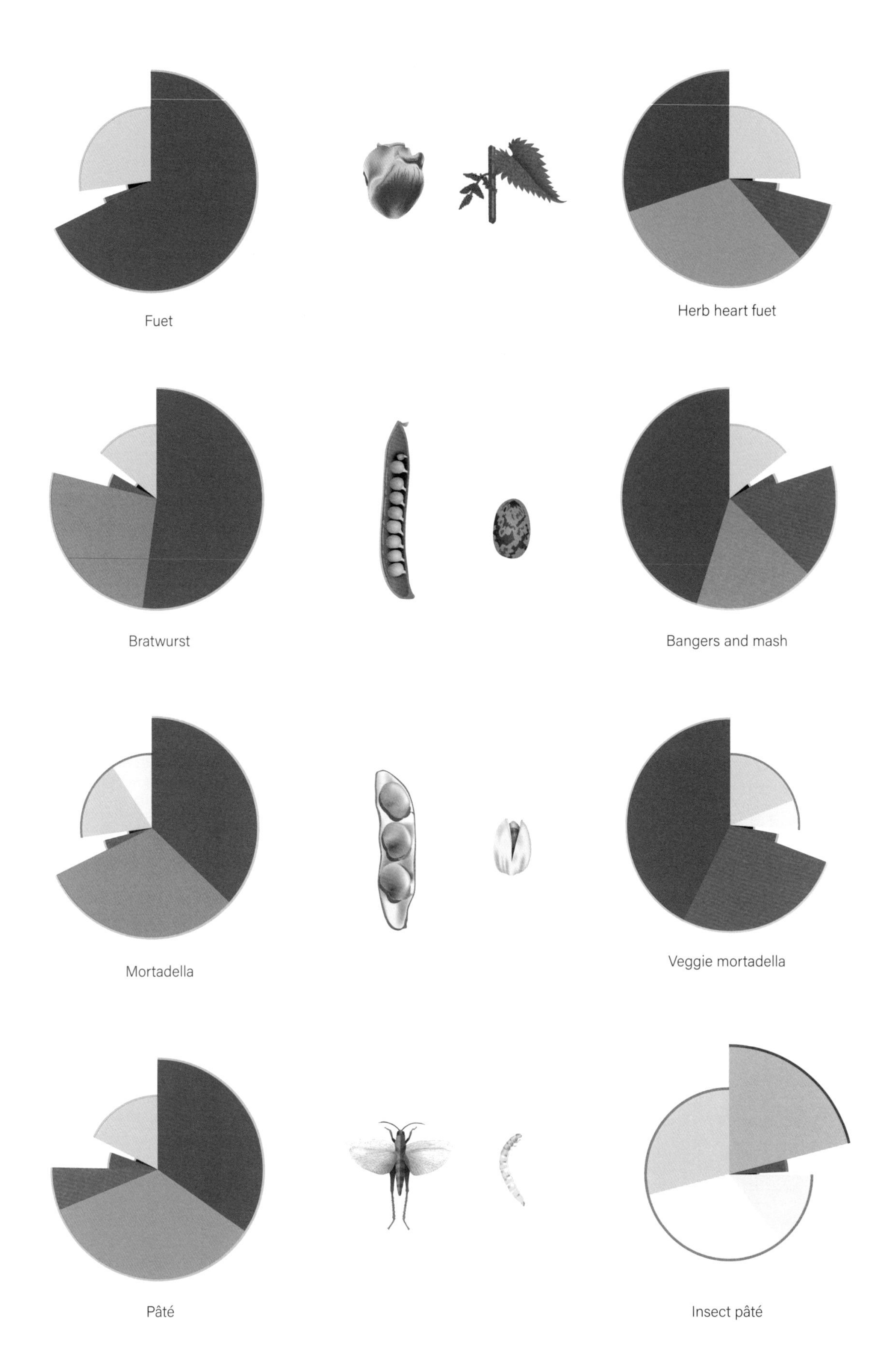
Fuet
Herb heart fuet
Bratwurst
Bangers and mash
Mortadella
Veggie mortadella
Pâté
Insect pâté

Sausage Matrix

Throughout this book, there have been clear divisions, categories, and techniques used to explain the elements and processes of the sausage. There is also further information on ingredients, glues, and environments. These formed the base of the construction of the sausage. All this information comes together in a matrix that is designed to explain and construct sausages schematically.

Diagram from the familiar old to the exciting new

The future sausages in this book have been constructed with the help of chef Gabriel Serero and butcher Herman Ter Weele. The ingredient chapter (p. 59) and the protein list (p. 63) are the basis for the mass of the future sausages. For example, the protein list shows which ingredients can be used as meat replacements, or which have high nutritional levels. All the sausages have been given a traditional type with a clear source of inspiration such as salami, pâté, boudin, bratwurst and mortadella and have been given additional future qualities such as being umami, hyper-nutritious or smokeable. These constraints bring fascinating results, some more realistic while others futuristic. The aim of the future sausage is to start reducing the amount of meat we consume from this moment on. That is why the sausages in this chapter are designed to be made with a butcher and with the same techniques and equipment used to fabricate the original versions. The following diagrams are designed to show visually what it means to replace the meat in a sausage, and especially how much more interesting the construction of a sausage can be compared to existing recipes. They are organised from left to right: from traditional to futuristic.

Meat mass
- Prime cut
- Offcut
- Offal
- Fat
- Blood

Non-meat mass
- Insects
- Legumes
- Vegetables
- Dairy and eggs
- Grains
- Seeds and nuts
- Fruits and berries
- Fish

Glue
- Animal protein
- Starch
- Hydrocolloid
- Fibre

Other
- Liquid
- Gel
- Spice
- Salt

Opposite page: the circles in the sausage diagrams are designed according to colour and size. Colour represents the type of ingredient (see legend above), whereas size indicates the purpose of an ingredient. The circles stand for, from large to small: mass, moisture, spice and salt; the rings represent the type of glue.

Following spread, left to right: the shift from traditional meaty sausages to future sausages, either redesigned or newly invented. Ingredients are further explained on pp. 150–51.

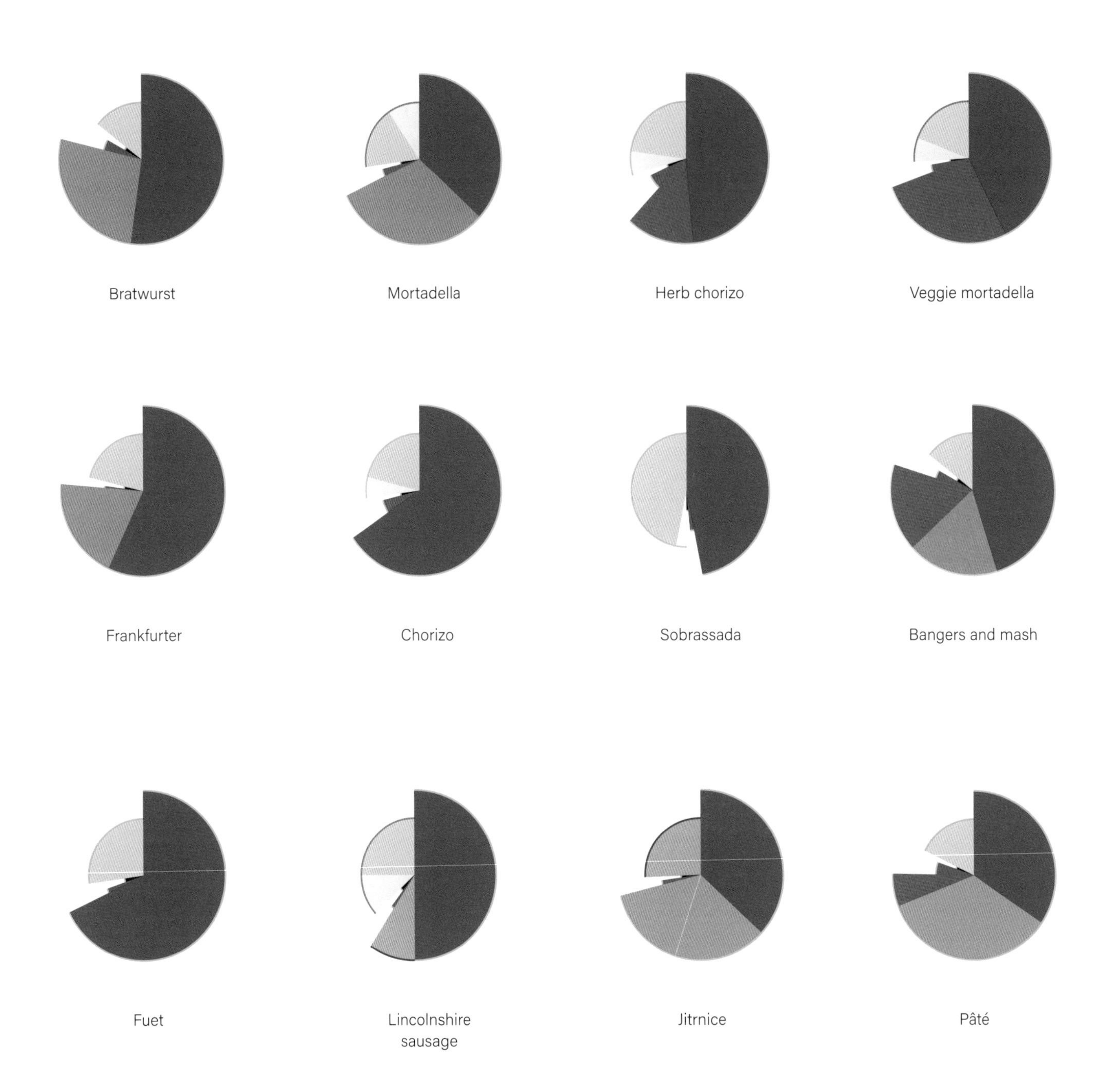

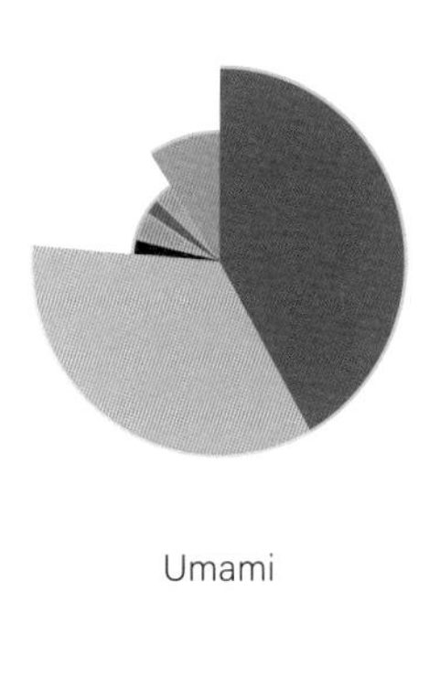

Umami

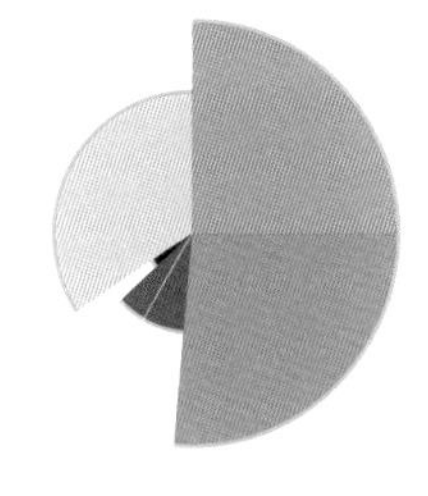

Hyper-nutritious

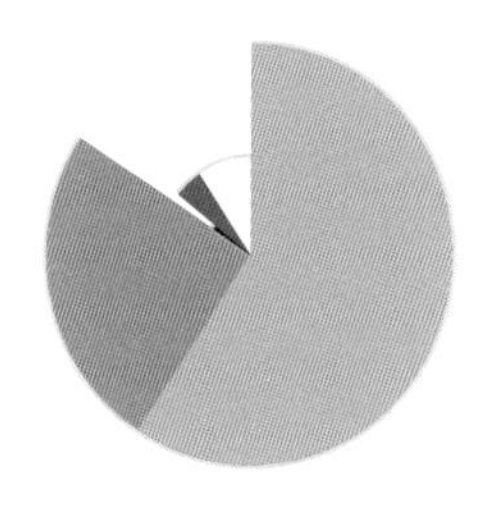

Polenta and apple

Apricot, carrot wiener

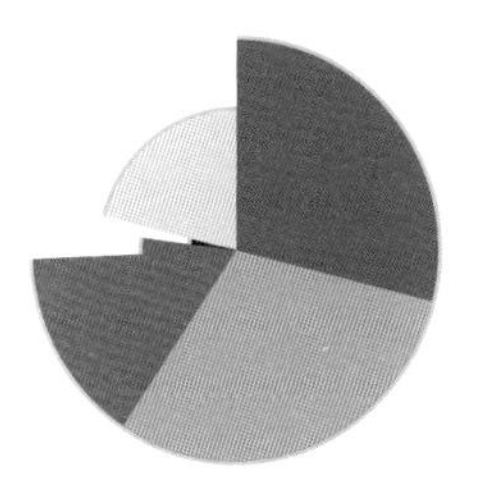

Liver sausage

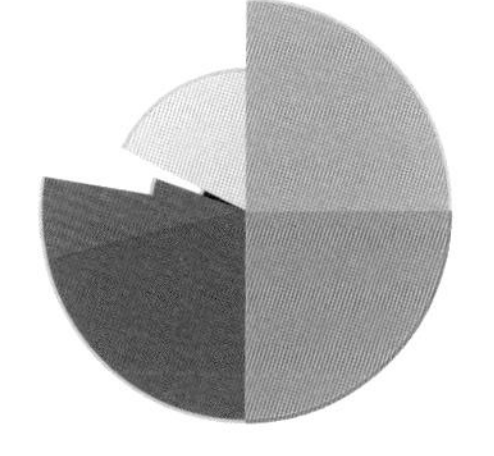

Liver and berry

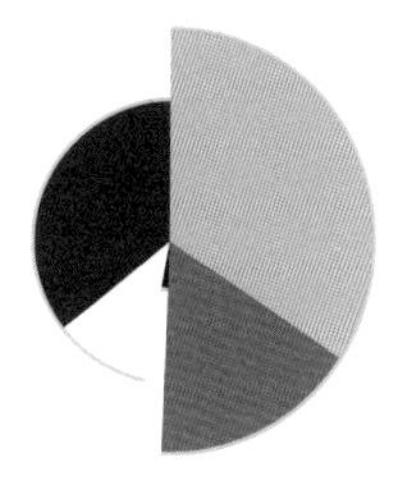

Blood sausage

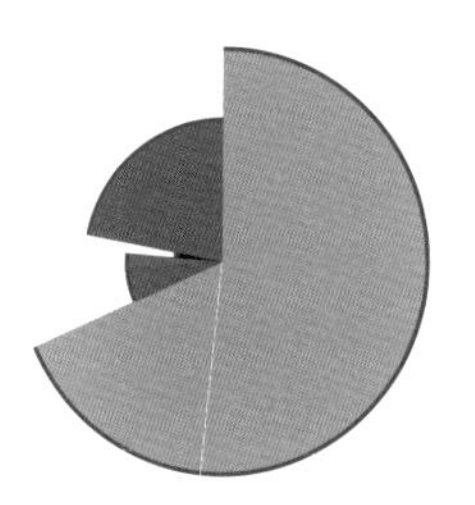

Green pea

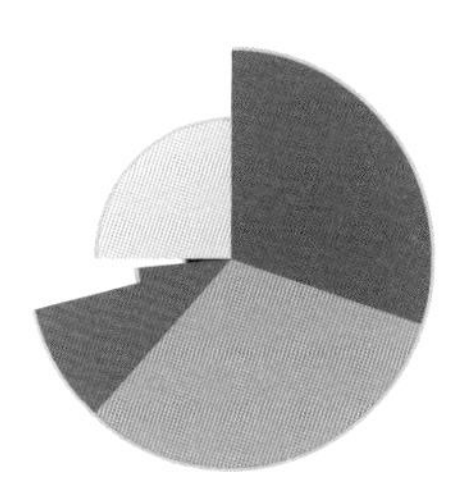

Herb and heart fuet

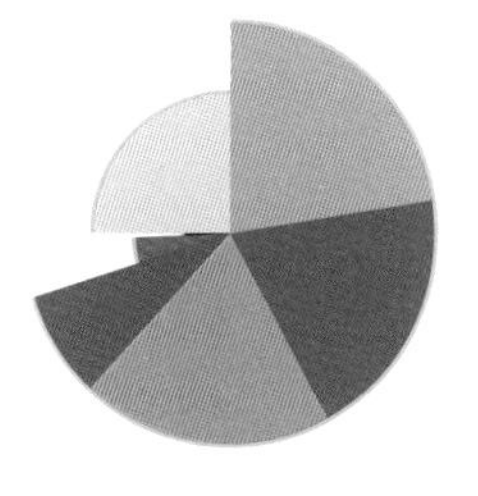

Sweetbread, flower banger

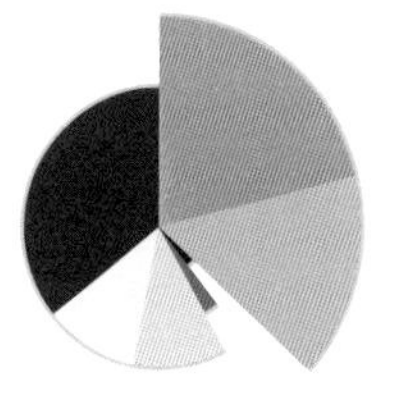

Apple boudin

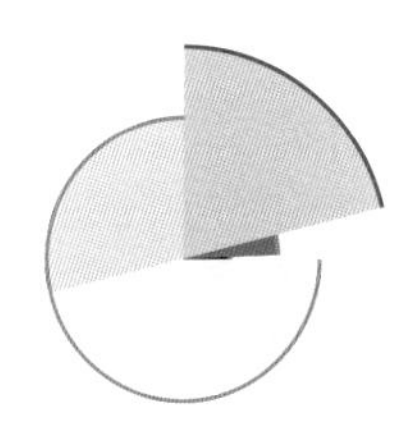

Insect pâté

Clockwise, from top left: carrot, apricot and coconut dried sausage; berry, date and almond dried sausage; insect salami; berry liver sausage; apple blood sausage; potato and pea fresh sausage; mortadella with asparagus.

Futurivore

Futurivore *noun*

An animal or person that eats food of both plant and animal origin with a future prospective in mind.

From Latin *futurum, futuri* n., posterity, future + *vorāre*, to eat greedily.

Sausages have always had strong character, both in flavour and appearance. They can be manly or rustic, or be more refined and humble; their flavour can be bold or subtle, and they can show off their ingredients proudly or conceal them when these are less conventional. All these properties lie in the design of a sausage.

The future sausages presented on the following pages have been selected according to their different character. Naturally, there are hundreds of different future sausage versions possible, just as there are hundreds of different sausages already in existence. The ones that feature in this chapter are those closely connected to the previously expounded sausage categories, and they represent some of the mouth-watering possibilities that lie ahead.

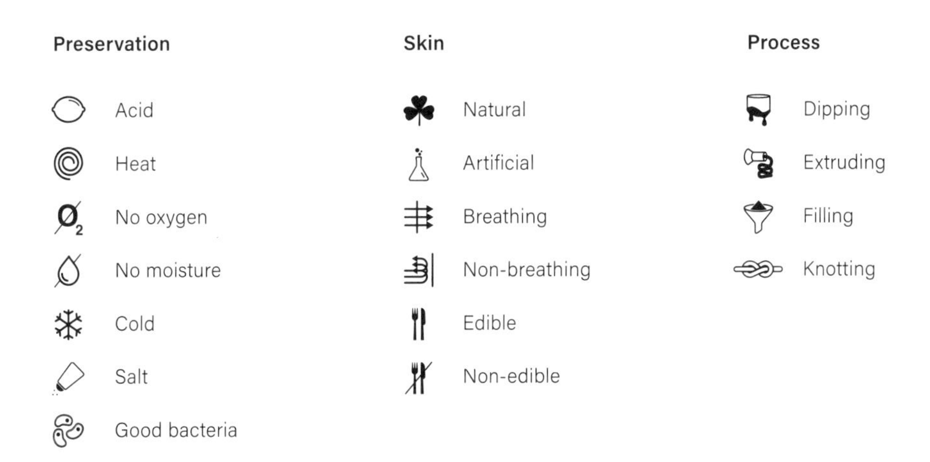

Above a legend to symbolize a preservation method, characteristic of the skin or the technique used to make the sausage. These symbols are used in the next chapter to explain the sausages.

Mortadella with vegetables

The mortadella, as explained in the cooked sausage section, is poached until cooked. This poaching ensures a bacteria-free environment and plays a major part in the preservation of the mortadella. The future version has a large quantity of cooked vegetables combined with the original mortadella mixture. Since the meat is poached to set, the blanched vegetables (which have been dusted with flour to absorb excess water) will require the same amount of time to cook, and will keep longer within the sausage. The vegetables have been selected for their fresh and sweet earthy aromas, to accompany the delicate taste of the mortadella meat. The mortadella can be sliced thinly and added to a sandwich, or cubed to be eaten as a snack. All in all, a good 20% less meat is consumed, while depth and flavour are added to the originally subtle flavour of the mortadella. The vegetables can easily be changed according to the season and the availability of local produce. The only set aspect is the water content of the ingredient. Of course, if an ingredient is too wet it can be dehydrated until it reaches a suitable state; think of semi-dried tomatoes or dried mushrooms – although obviously not all flavours suit the mild one of the mortadella.

Composition: mortadella (scale 1:2)

40% Pork meat
28% Broccoli, carrot
18% Pork back fat
8% Water
4% Pistachios, pepper, mace, nutmeg
2% Salt

Glue: myosin

Skin

Process

Preservation

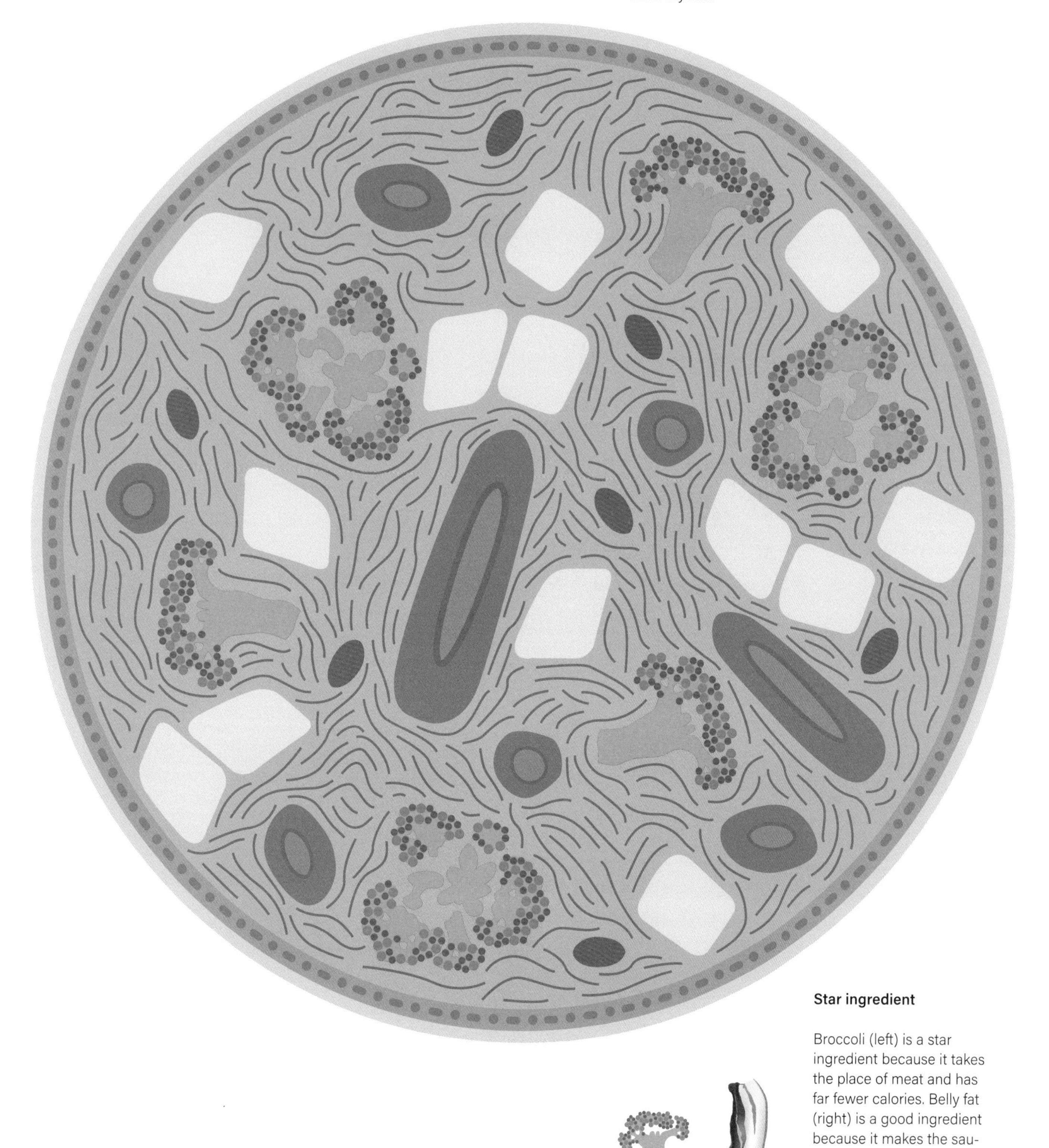

Star ingredient

Broccoli (left) is a star ingredient because it takes the place of meat and has far fewer calories. Belly fat (right) is a good ingredient because it makes the sausage creamy and takes the place of prime cuts.

Mortadella with vegetables is a sausage with a pleasant, mild flavour, which goes well with the delicate flavour of vegetables.

Bangers and mash & apricot and carrot & pea and chickpea

Bangers, or fresh sausages, are available in many different varieties, particularly as they are easy to make and are a favourite for both barbecues and breakfasts. Bangers are usually made in two different sizes. One version, made in pork intestines, is around 30 millimetres in diameter and 120 millimetres in length; the smaller version, made in lamb casing, is 20 millimetres in diameter and often around 80 millimetres in length. Because it is a fresh sausage, preservation is not a very important issue; it is mostly about the size of the chunks, the moisture content, and the flavour. For the future bangers, the skin remains an animal intestine, but can easily be replaced with a cellulose version. If an intestinal casing has been kept here, it is because the idea is not to ban meat completely. The advantage of this skin is the taste and delicacy it brings to the sausage.

The bangers and mash sausage has been filled with a pea puree and mashed potato puree, as well as 60 per cent of the traditional meat mixture. The sausage fries nicely in the pan and holds within it the delicious flavours of the British national dish bangers and mash. This recipe easily allows one to use up to 40 per cent less meat, and even add a small percentage of chia seeds, which are high in antioxidants. The flavour profile has thus been lifted and the protein content has remained high. As a production method, the extruder can be filled with a marbled mixture of the three ingredient pastes and extruded at the same time into the casing.

Pea banger & apricot and carrot banger contain no meat at all yet comprise a fascinating mix of flavours and nutrition. The pea banger is made with peas, chickpeas, edamame beans and lots of herbs. The sausage thus contains fewer calories but remains rich in protein because of the chickpeas, and has a higher nutritional value than a traditional banger owing to the green peas. The apricot and carrot sausage is lightly sweet, savoury and nutty in flavour, with a sharp, fresh punch of ginger. It is held together by the protein-rich cashew flour, and is lovely to grill, especially over an open fire. It can be enjoyed in many ways, for example in a sandwich with some salty cheese.

Composition: bangers and mash, pea banger and carrot banger (scale 1:1)

45%	Lean pork meat	35%	Chickpeas	35%	Carrot
18%	Green pea puree	28%	Green peas	30%	Almond flour
18%	Potato puree	18%	Edamame beans	25%	Apricot
15%	Pork belly fat	15%	Sesame paste	6%	Ginger
3%	Chia seeds	3%	Herbs and flowers	3%	Lemon juice
1%	Salt	1%	Salt	1%	Salt

Glue: myosin

Glue: fibre

Glue: fibre

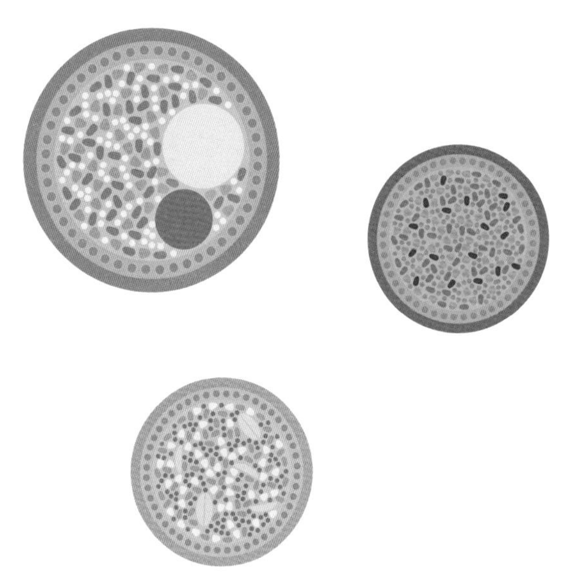

Star ingredient

Chicories (left) can give a subtle bitterness when added to the green pea sausage. Almonds (middle) give great flavour but, even better, bind the sausage and give it proteins. Green peas (right) add freshness, sweetness, and nutrition.

Two types of vegetarian sausage, inspired by the British national dish bangers and mash, consisting of mashed potatoes, peas, flowers and herbs and carrots, ginger and apricots.

Heart fuet

When asking a butcher for his or her favourite cut of meat, I was surprised to hear pig's heart mentioned several times. Upon further investigation, it turned out that heart meat is beautifully lean and has muscle-like properties. In the heart fuet, this ingredient is used to lower the fat content and intensify the meaty tanginess of a good dried sausage. The herbs both give an intense savoury note, because they season the meat during the drying process, and add extra preservation qualities owing to their preserving properties. This sausage can be made with the same techniques as dried sausages and enjoyed in the same way.

Composition: herb and heart fuet (scale 1:1)

30% Lean pork meat
30% Heart meat
25% Pork belly fat
10% Herbs
3% Spice: fennel seeds
2% Salt

Glue: myosin

Skin

Process

Preservation

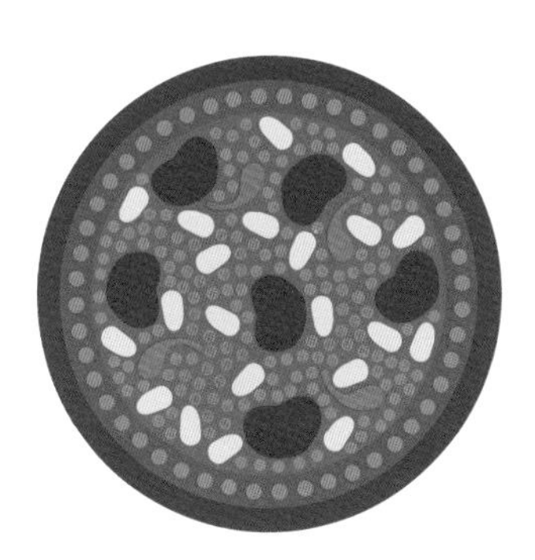

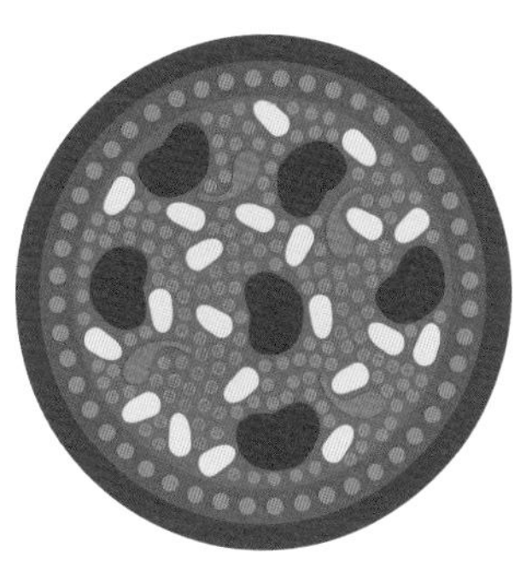

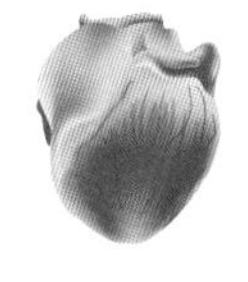

Star ingredient

The heart (left) and nettles (right) are the star ingredients: the heart gives a tangy, savoury addition, and the nettles provide a herbal, grassy note to the sausage.

A sausage designed to accompany cheese and fruit, with the addition of the lean single-muscle meat of the heart, which cures evenly with the rest of the sausage.

Fruit salami

The fruit salami is entirely made out of fresh and dried fruits with a structural base of hazelnut and almond flour. The texture comes from pieces of almonds and flowers. This sausage contains no meat, thus reducing the meat content by 100 per cent. The fruit salami is made with the same machines as other dried sausages, but is dried faster at a slightly higher temperature, since it does not need the same slow curing time as a meat-based dried sausage. The tangy, slightly sweet flavour profile is calculated to best accompany dried meats, cheeses, and other foods that are traditionally served with an aperitif, but it can also be enjoyed simply sliced with some cheese on toast. The sausage is held together by the fibres of the semi-dried fruit. Depending on how long it has been drying, the center can be juicy and moist, or chewier and drier.

Composition: fruit salami (scale 1:1)

30% Forest fruits
30% Dried fruits
20% Almond flour
15% Almonds
4% Flowers
1% Salt

Glue: pectin

Skin

Process

Preservation

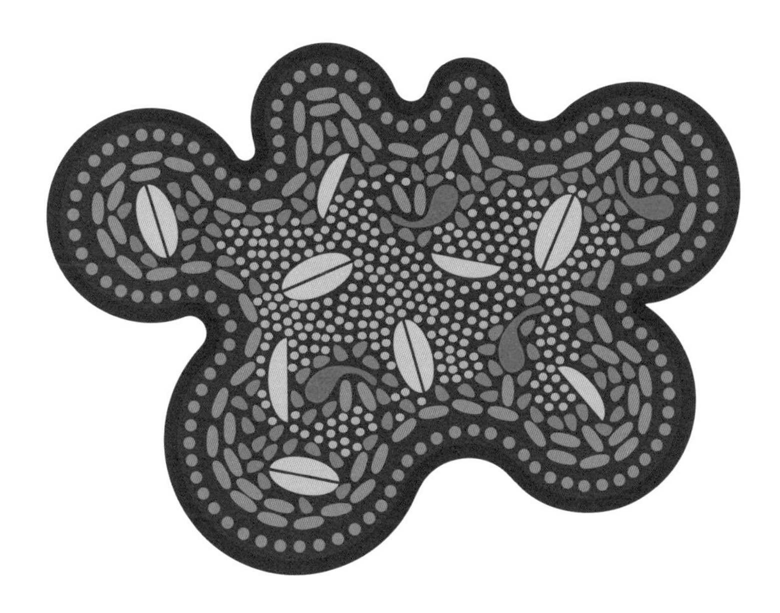

Star ingredient

Almonds (left) and nasturtium (right) are the star ingredients: almonds because of the proteins they contain and nasturtium because it gives an extra floral note that upgrades the flavour profile.

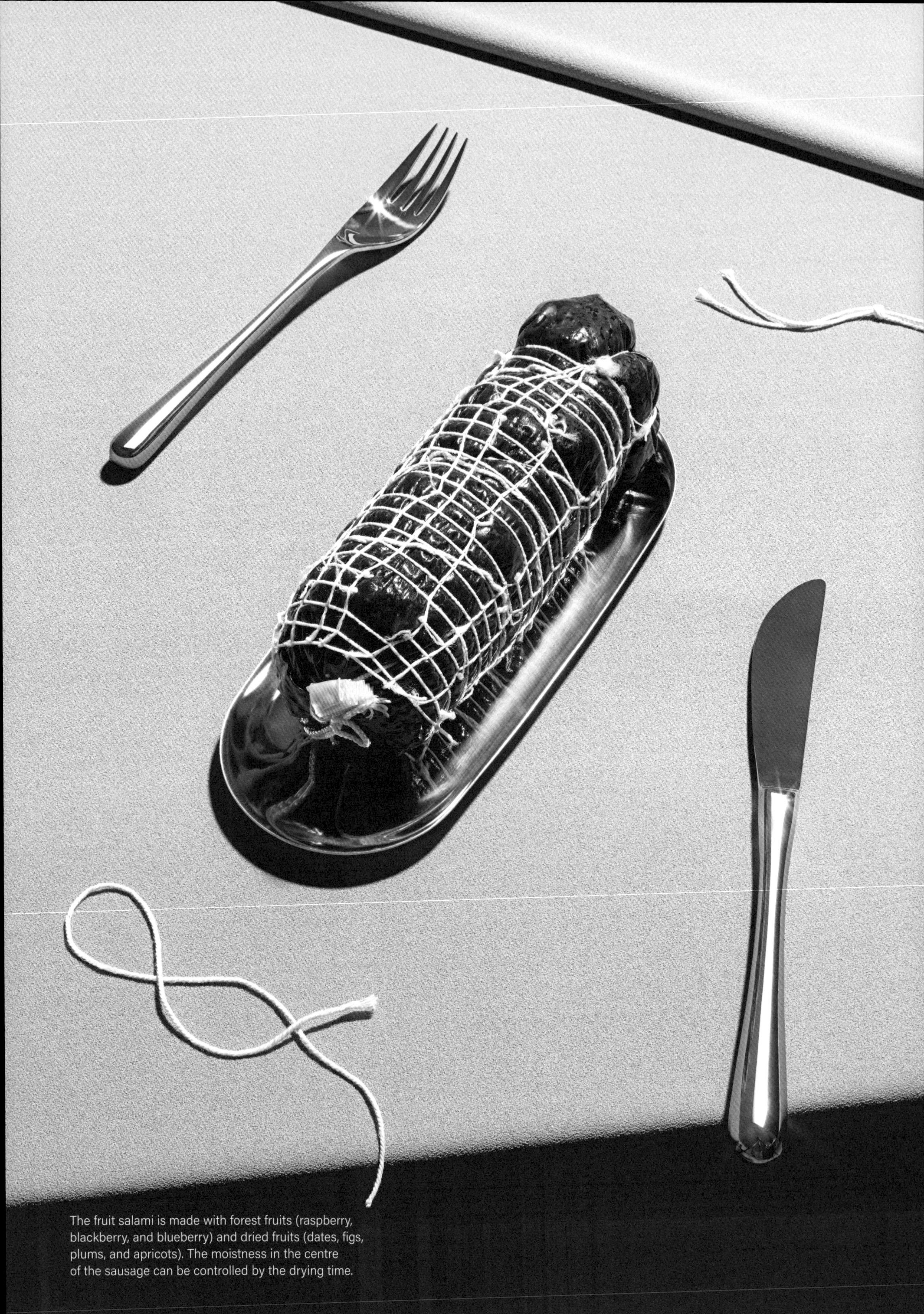

The fruit salami is made with forest fruits (raspberry, blackberry, and blueberry) and dried fruits (dates, figs, plums, and apricots). The moistness in the centre of the sausage can be controlled by the drying time.

Liver and berry

Traditionally, many types of meats and especially liver-based foods are eaten with a fruit-based accompaniment. This is the inspiration for the liver and berry sausage, which substitutes part of the meat content with a gel made from raspberries. The sausage is already based on a model that uses offal and generally consists of a small percentage of meat. The addition of fruit gives a more full-bodied flavour, because the raspberry has a surprisingly roasted note which matches with the liver perfectly. Moreover, the liver has herbal and green notes that make its bond to the raspberry even stronger.

Composition: liver and berry (scale 1:1)

25% Pork liver
25% Raspberry gel
18% Pork shoulder
10% Onion
15% Pork belly fat
3% Spices: mace, nutmeg and black pepper
1% Salt

Glue: myosin and pectin

Skin

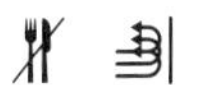

Process

Preservation

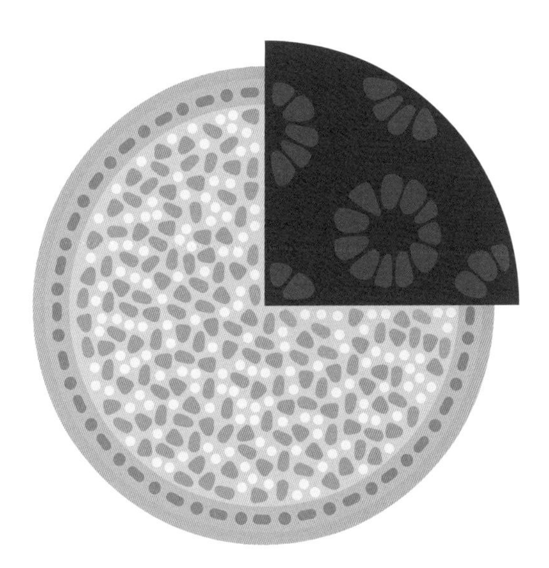

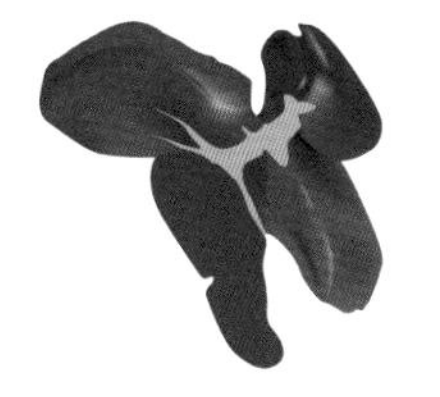

Star ingredient

Liver (left) remains a star ingredient because of its delicate earthy flavour, and the raspberry jelly is a star ingredient because of its complementary fresh and sweet note, which matches the liver flavour perfectly.

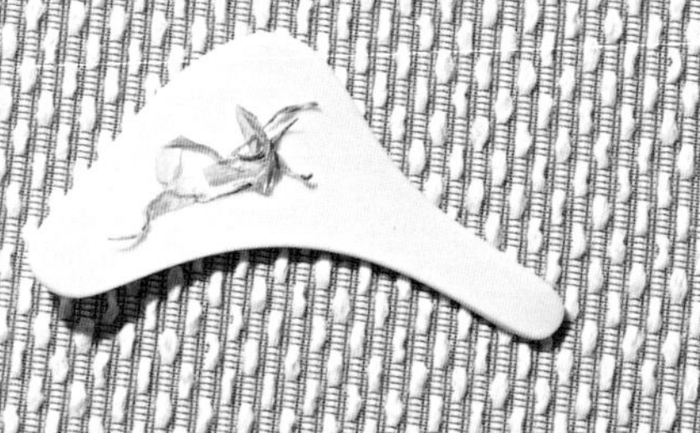

Traditionally, many types of meats and especially liver-based foods are eaten with a fruit-based accompaniment. Here, the mild liver is combined with a tangy and sweet raspberry gel.

Apple boudin

Blood sausages have always been either highly praised or avoided at all costs. The latter has most likely nothing to do with the taste, but merely with the idea of eating blood. However, this versatile animal product is exceptionally pleasant and mild in taste, and is often combined with apples for a note of tart sweetness. The future version is made according to the traditional Dutch version of baked blood sausage with barley, with the addition of chocolate and almonds and an apple compote core thickened with agar, which has a high resistance to heat. The slices of apple boudin can be fried as normal in a pan. Due to the agar, the apple will melt slowly and remain in the centre of the *boudin*, giving it a caramelised apple crust and soft apple centre. A fully vegetarian version can be imagined to be made of a polenta-based body with a Granny Smith apple gel core and a black curry spice coating. The *polenta* apple sausage can be sliced and fried in a pan similarly to the apple boudin, and has a pleasantly mild polenta flavour combined with tart apple and a smoky curry seasoning.

Composition: apple boudin (scale 1:1)

35% Blood
18% Golden apple
15% Barley
10% Pork belly fat
8% Milk
8% Roasted almonds
2% Cacao powder
2% Spices: white pepper and clove
2% Salt

Glue: myosin and hydrocolloid

Skin

Process

Preservation

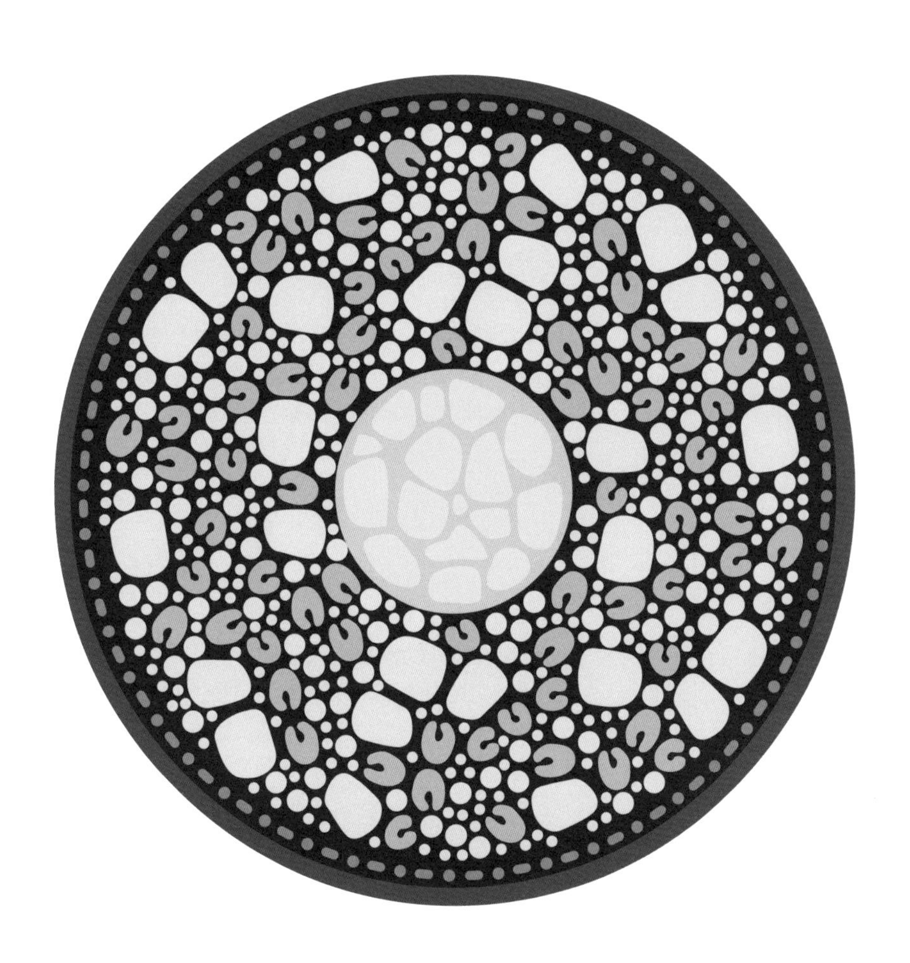

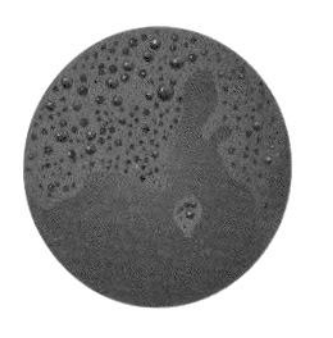

Star ingredient

Pork blood (left) remains the star ingredient because of its high nutritional value and its abundance when slaughtering animals. The apple is there to create a more approachable product. The wheat (right) adds to the texture and bulk, while the apple core lends a sweet and tangy flavour.

Blood sausage is traditionally eaten with apple, and here the two ingredients are combined in one sausage. The core of apple gel is set with agar, a substance with a high melting point, which keeps the gel from running during cooking.

Insect pâté

The insect pâté is made from mealworm flour. Insects are exceptionally high in proteins, and thus highly suitable as a protein replacement. The nuttiness they bring is complimented here by the oriental seasoning. Insect pâté is the furthest away from the classic design of a sausage, but nonetheless it is edible in a casing. This sausage is the most mysterious, and has a certain chic quality that comes from the soft and tactile wax casing. The casing can be cut open and the pâté scooped out. The first encounter with this sausage is confusing yet intriguing, much like the idea of eating insects. Hence the mission of this futuristic food item is to be a gateway sausage towards the path of embracing insect consumption. Research has shown that to overcome reluctance towards an ingredient or flavour, one should be exposed to it several separate times in a comfortable setting: a method that has proven to be very effective.

Composition: insect pâté (scale 1:1)

35% Carrot juice
30% Pecan flour
20% Insect flour
10% Capers, jalapeño, flowers and pecan
3% Oriental cumin
2% Salt

Glue: fibre

Skin

Process

Preservation

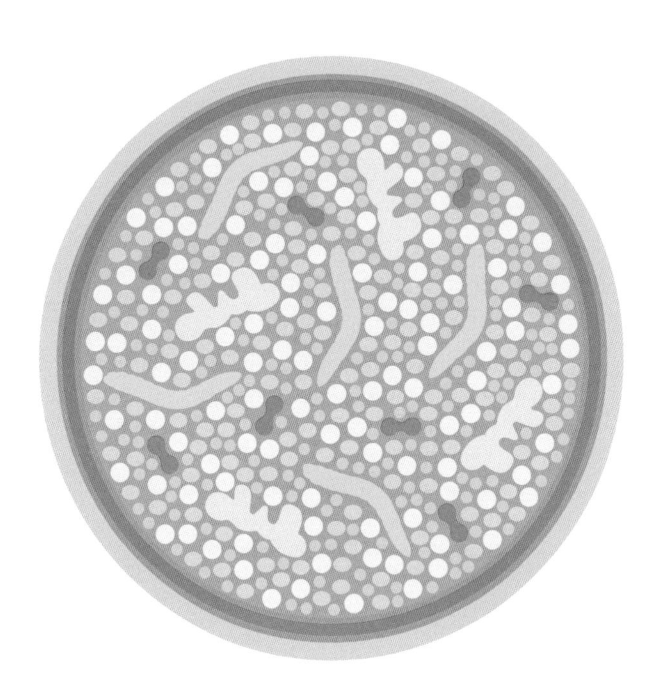

Star ingredient

Insects are the star ingredient in this sausage because of their novelty and nutritional values. The protein level in insects is extremely high, making them a suitable meat substitute.

Nutty with the taste of insect flour and a oriental cumin infusion, the insect pâté is dipped in an airtight beeswax casing to prevent it from discolouring and to increase its shelf life.

The Psychology of Disgust and the Desire for Delicacy

When looking back on the research for this book, I found that the most common barrier that needed to be crossed was a mental one. The machines, techniques and ingredients for the creation of future sausages are already in place and available, but we avoid certain foods simply because we do not like the idea of them, not because they are bad for us or are unobtainable. This avoidance is called the *Psychology of Disgust*. Dr Paul Rozin, professor of psychology at the University of Pennsylvania and an expert on human disgust, says that the barriers against certain foods are psychological: disgust is a revulsion response, a basic biological motivational system. It protects us from getting sick, since – as Darwin posited – food that tastes bad is also bad for one's health. Today, our reluctance is still controlled by this conditioned defence mechanism, which is often no longer biological but is created by our culture. Or as Rozin puts it: 'Disgust evolves culturally, and developed evolutionarily from a system to protect the body from harm to a system to protect the soul from harm.'

But if it is edible, delicious and nutritious, why won't we eat it?

We ate efficiently in the past, and our current culture is making us pickier and less efficient. This book was therefore meant to break open an established food item and make it fresh with new combinations, yet during the research process I found that we have constantly been creating new foods, rather than rediscovering them. Maize has gone through many transformations to become the juicy sweet grain it is today, and the bitter self-defence mechanism of many plants that are now our delicate salads have been developed by crossbreeding and agriculture. Inspired by the drive of our ancestors to make as many different foods as possible, we need to continue on this path and be open to new creations.

The future sausage is a metaphor for the possibilities that lie ahead; it should revive creativity and curiosity in our eating habits. Let us not be wary of food, but explore and appreciate it instead.

Glossary

Sausage Matrix
Bibliography
Biographies
Colophon

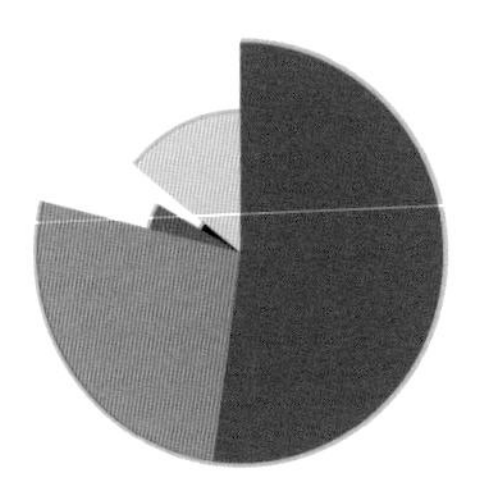

Bratwurst
53% Pork shoulder
27% Lean veal offcuts
14% Pork fat
4% Spice
2% Salt
Casing: intestines
Glue: myosin
Preservation: cold

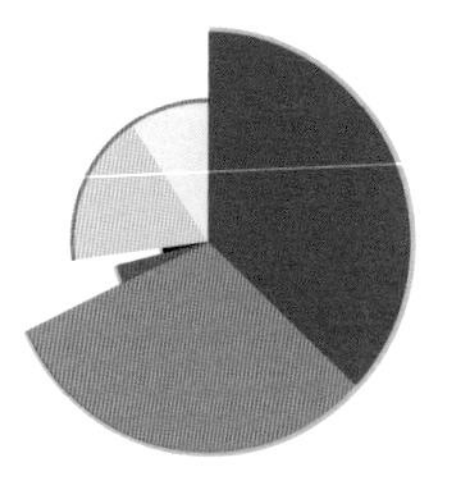

Mortadella
38% Pork shoulder
32% Pork trimmings
15% Fat
8% Water
5% Pistachios, clove, pepper, mace
2% Salt
Casing: collagen, ox runner
Glue: myosin, hydrocolloid
Preservation: no oxygen, heat

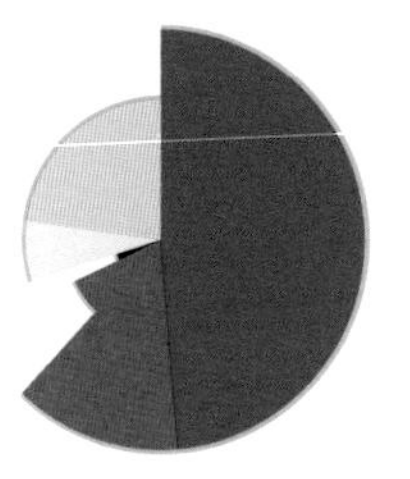

Herb chorizo
48% Pork shoulder
14% Herbs
22% Fat
9% Vinegar
5% Chilli, mace, garlic
2% Salt
Casing: intestines
Glue: myosin
Preservation: no moisture, good bacteria

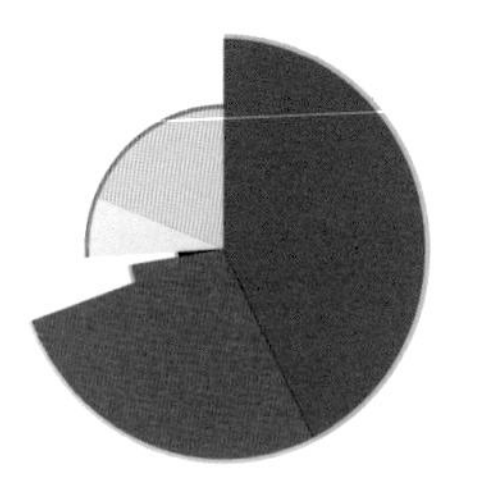

Veggie mortadella
40% Pork meat
28% Vegetables: broccoli, carrot, romanesco, cauliflower
18% Pork back fat
8% Water
4% Pistachios, pepper, mace, nutmeg
2% Salt
Casing: collagen, ox runner
Glue: myosin, hydrocolloid
Preservation: heat

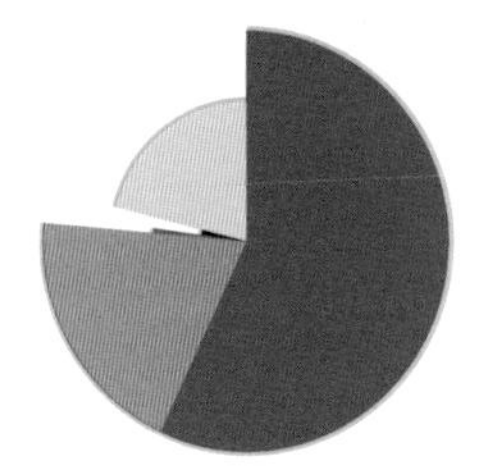

Frankfurter
58% Beef
19% Pork trimmings
20% Pork fat
2% Salt
1% Spice
Casing: cellulose
Glue: myosin
Preservation: no oxygen, heat

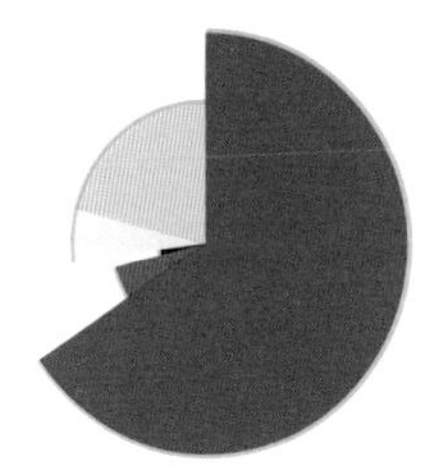

Chorizo
65% Pork shoulder
20% Fat
8% Vinegar
5% Chilli, paprika, garlic
2% Salt
Casing: intestines
Glue: myosin
Preservation: no moisture, good bacteria

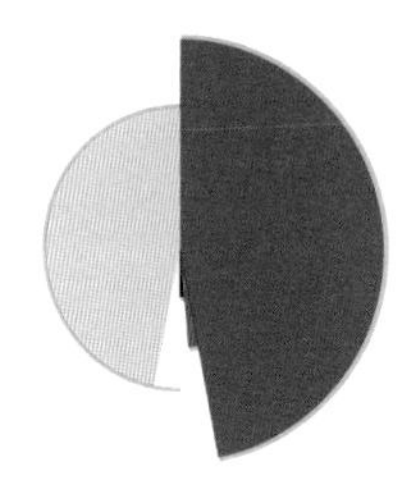

Sobrassada
46% Lean pork
46% Pork fat
4% Goat's cheese
2% Salt
2% Spice
Casing: intestines, bladder, stomach
Glue: myosin
Preservation: good bacteria

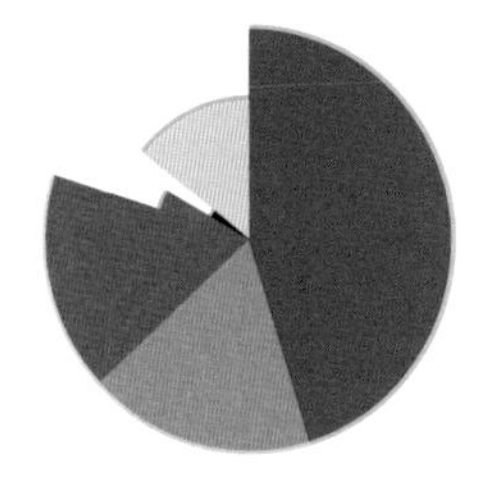

Bangers and mash
45% Lean pork
18% Green-pea puree
18% Potato puree
15% Pork belly and back fat
3% Chia seeds
2% Salt
Casing: intestines
Glue: myosin, starch
Preservation: cold

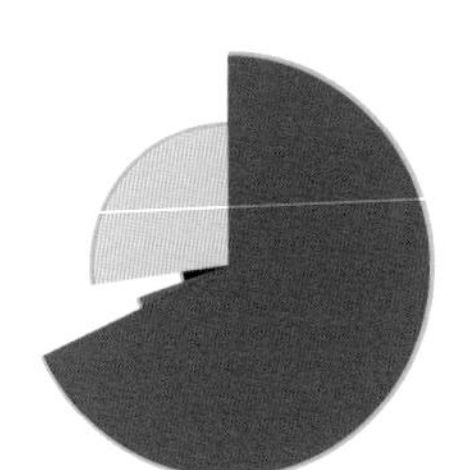

Fuet
67% Lean pork
28% Back fat
3% Salt
2% Fennel, pepper, garlic
Casing: intestines
Glue: myosin
Preservation: no moisture, good bacteria

Lincolnshire sausage
50% Pork shoulder
10% Breadcrumbs
25% Pork belly
12% Water
2% Salt
1% Sage, pepper
Casing: intestines
Glue: myosin, fibre, hydrocolloid
Preservation: heat

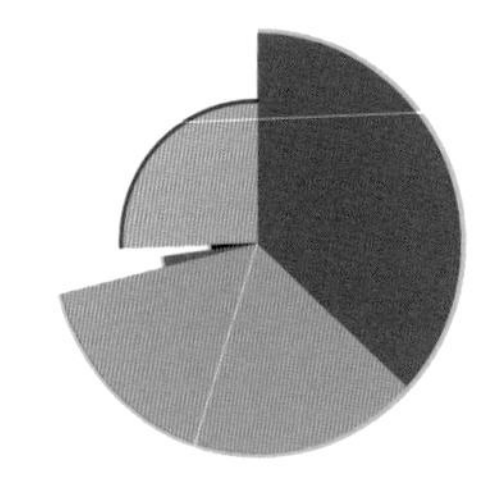

Jitrnice
38% Pork meat
14% Offal (lungs, Spleen, stomach)
18% Liver
26% Soaked Breadcrumbs
2% Salt
2% Spice
Casing: intestines
Glue: myosin, fibre
Preservation: heat

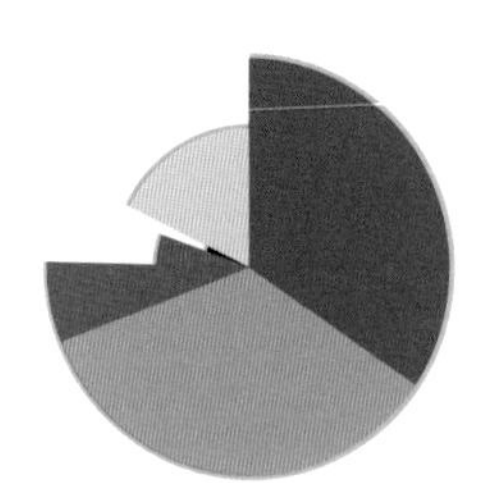

Pâté
35% Pork shoulder
35% Pork liver
8% Onions
15% Pork belly fat
5% Spice
2% Salt
Casing: ox runner, pork fat, nylon
Glue: myosin
Preservation: heat, no oxygen

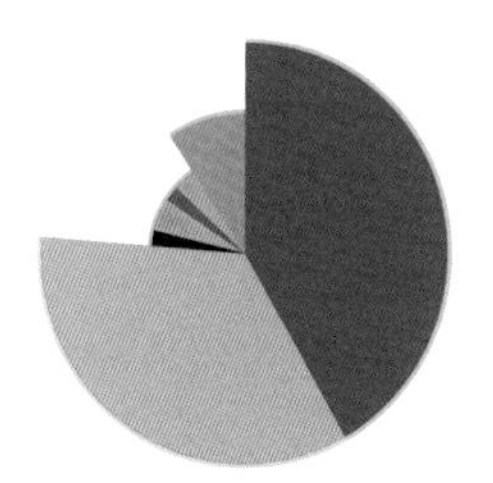

Umami
43% Beef
35% Barley
10% Miso
5% Green olives
3% Preserved lemons
2% Salt: anchovies
2% Spice: black curry
Casing: algae
Glue: myosin, starch
Preservation: cold

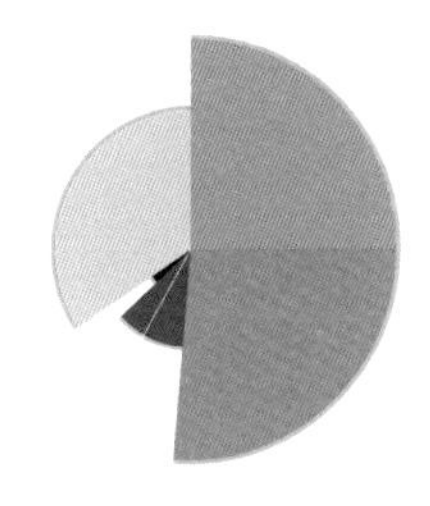

Hyper nutritious
25% Beef liver
27% Green peas
35% Pork belly
7% Pistachios
4% Chia seeds
2% Salt
Casing: collagen, nylon, fabric
Glue: myosin
Preservation: heat

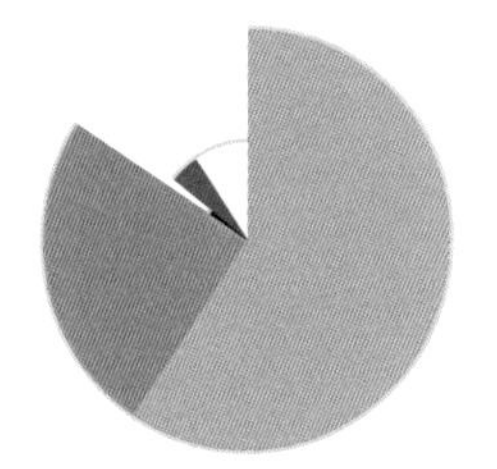

Polenta and apple
60% Polenta
25% Granny Smith apples
10% Parmesan cheese
4% Black curry
1% Salt
Casing: intestines, cellulose
Glue: starch
Preservation: cold

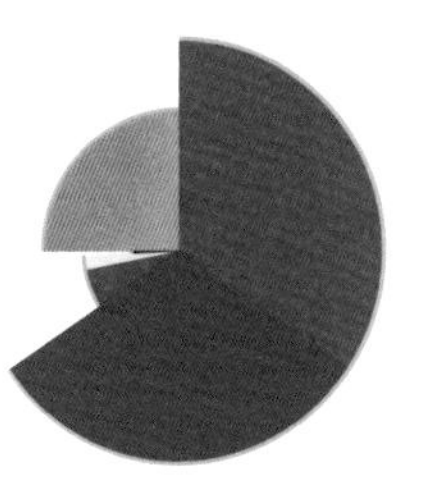

Apricot and carrot wiener
35% Carrot
30% Almond flour
25% Apricots
6% Ginger
3% Lemon juice
1% Salt
Casing: collagen
Glue: fibre
Preservation: cold

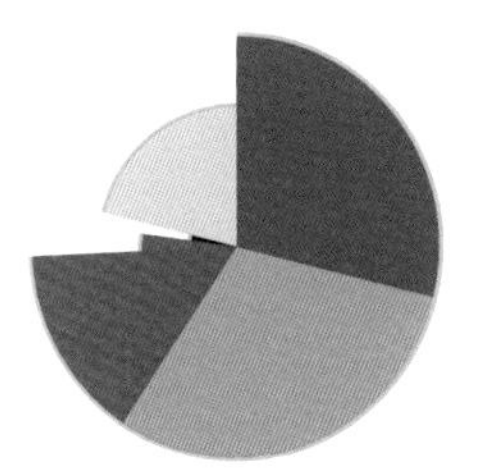

Liver sausage
30% Pork shoulder
30% Pork liver
14% Onions
22% Pork belly fat
4% Spices
2% Salt
Casing: ox runner, collagen, nylon
Glue: myosin
Preservation: heat, no oxygen

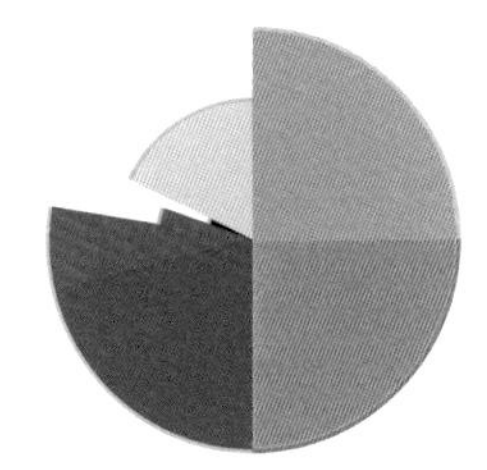

Liver and berry
25% Pork liver
25% Raspberry gel
20% Pork shoulder
10% Onions
15% Pork belly fat
3% Spices
2% Salt
Casing: ox runner, collagen, nylon
Glue: myosin, hydrocolloid
Preservation: heat, no oxygen

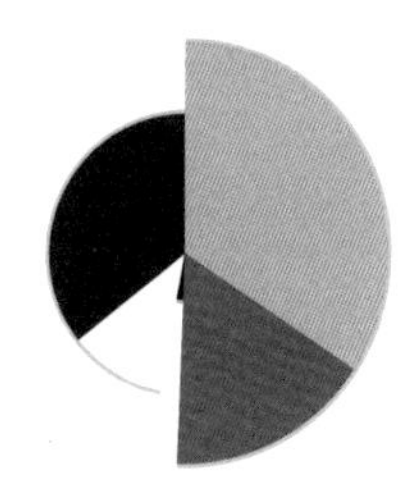

Blood sausage
36% Oats
15% Onions
36% Blood
10% Cow's milk
3% Salt
Casing: collagen, ox runner
Glue: myosin, hydrocolloid
Preservation: heat

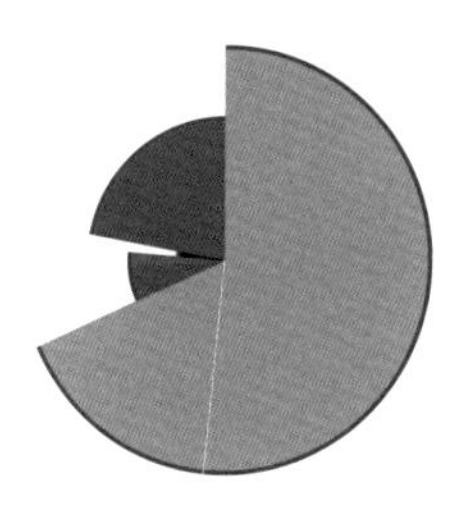

Green pea
50% Green peas
17% Shredded tempeh
21% Sesame paste
5% Walnuts
3% Spice
Casing: nylon
Glue: fibre
Preservation: cooking

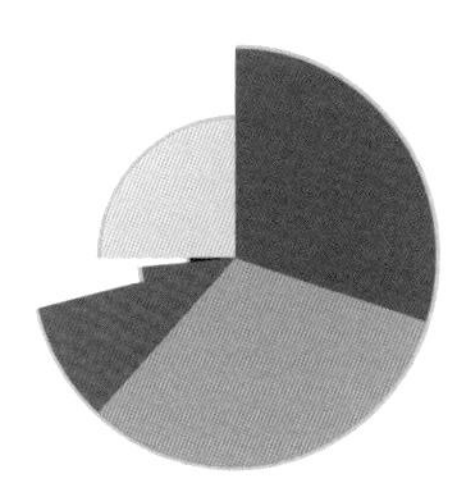

Herb and heart fuet
30% Lean pork meat
30% Pig's heart meat
10% Herbs
25% Pork belly and back fat
3% Fennel seeds
2% Salt
Casing: intestines, collagen
Glue: myosin
Preservation: no moisture, good bacteria

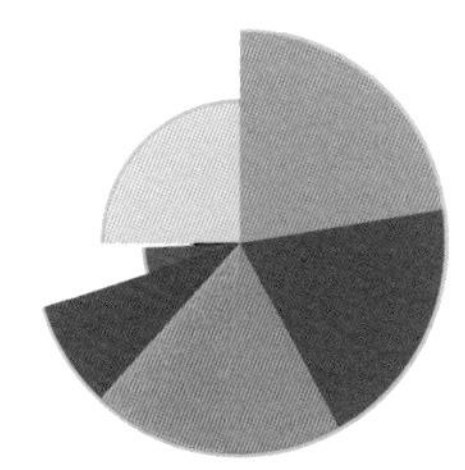

Sweetbread and flower banger
22% Sweetbread
20% Pork shoulder
10% Pork trimmings
10% Apple
7% Hazelnuts
25% Pork belly
5% Dried flowers
1% Salt
Casing: intestines
Glue: myosin
Preservation: cold

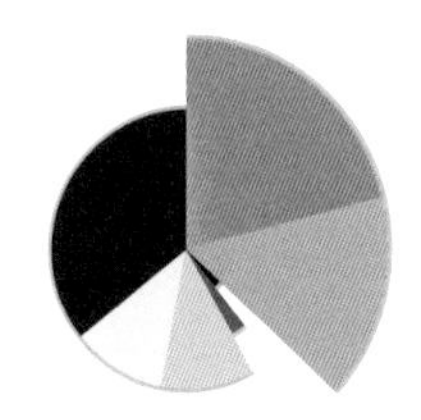

Apple boudin
16% Barley
18% Apple
36% Pig's blood
12% Pork belly fat
10% Milk
2% Salt
2% Spices
Casing: collagen, ox runner
Glue: myosin, hydrocolloid
Preservation: heat

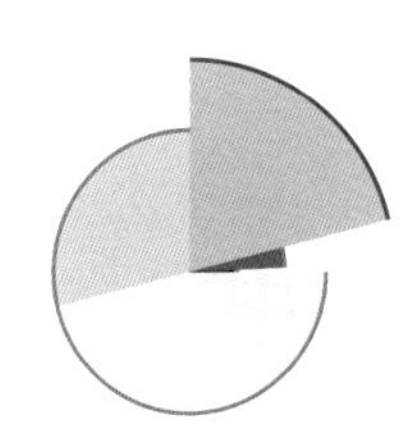

Insect pâté
20% Insect flour
30% Ewe's milk cream
30% Butter
15% Madeira wine
4% Tonka bean
1% Salt
Casing: beeswax
Glue: hydrocolloid, fibre
Preservation: no oxygen

Bibliography

A selection of sourses consulted for the writing of this book:

De smaakbijbel, Niki Segnit, 2011.
Edible insects: Future prospects for food and feed security by the Food and Agriculture Organization of the United Nations (FAO), 2013.
Modernist Cuisine: The Art and Science of Cooking, Nathan Myhrvold, Chris Young and Maxime Bilet, 2011.
Noma, Rene Redzepi, 2011.
Offal, A global history, Nina Edwards, 2013.
On Food and Cooking: The Science and Lore of the Kitchen, Harold McGee, 2004.
Over Worst, Meneer Wateetons & Sjoerd Mulder, 2011.
Sausage, Nichola Fletcher, 2012.

American Chemical Society, Cottingham, Katie, https://www.acs.org/content/acs/en/pressroom/newsreleases/2016/august/edible-food-packaging-made-from-milk-proteins-video.html (accessed November 22, 2016)
David Ideas, David Edwards, http://www.davidideas.com/(accessed February 25, 2015)
Food and Agriculture Organization of the United Nations, http://www.fao.org/docrep/meeting/024/Mc937e.pdf (accessed February 23, 2015)
Food and Agriculture Organization of the United Nations, http://www.fao.org/docrep/005/y4351e/y4351e0c.htm (accessed February 20, 2016)
Food pairing, https://www.foodpairing.com/en/home (accessed September 24, 2016)
Gustav Heess, http://gustavheess.de/index.php?option=com_content&view=article&id=286%3Akuerbiskernoel&catid=65%3Alebensmittel&Itemid=126&lang=en (accessed July 13, 2015)
Journal of Biological Chemistry, D. Breese Jones and Charles E. F. Gersdorff, http://www.jbc.org/content/75/1/213.full.pdf (accessed February 02, 2016)
Online Entymology Dictionary, Harper, Douglas, http://www.etymonline.com/index php?allowed_in_frame=0&search=sausage (accessed October 13, 2016)
Self Nutrition Data, http://nutritiondata.self.com/(accessed October 13, 2016)
The American Association of Cereal Chemists, Michael J. Gooding, http://aaccipublications.aaccnet.org/doi/abs/10.1094/9781891127557.002 (accessed March 23, 2016)

Biographies

Carolien Niebling (b. 1984) is a designer and researcher who specializes in food-related projects and lives and works in Lausanne, CH. Graduated ECAL Master Product Design in 2014. Winner of the Grand Prix, Design Parade at Villa Noailles, Hyères, 2017.
www.carolienniebling.com

Herman ter Weele (b. 1975) is an award-winning butcher at Slagerij G. ter Weele & Zoon in Oene, NL.
www.slagerijterweele.nl

Gabriel Serero (b. 1977) is a Swiss chef who specializes in food-texture at Emotion Food Company, Lausanne, CH.
www.emotionfood.ch

Jonas Marguet (b. 1982) is a still life photographer who works and lives in Lausanne, CH. Graduated ECAL Bachelor Photography in 2008 and Master Art Direction in 2011.
www.jonasmarguet.com

Emile Barret (b. 1989) is a photographer and artist, based in Paris, FR, and works all over the world. Graduated ECAL Bachelor Photography with Honours in 2012.
www.emilebarret.com

Noortje Knulst (b. 1987) is a still life photographer who works and lives in Amsterdam, NL. Graduated ECAL Master Art Direction with Honours in 2014.
www.noortjeknulst.com

Helge Hjorth Bentsen (b. 1988) is a graphic designer who works and lives in Oslo, NO. Graduated ECAL Bachelor Graphic Design with Honours in 2012.
www.helgebentsen.com

The Sausage of the Future
A research project by Carolien Niebling conducted at ECAL

Concept and editor Carolien Niebling, www.carolienniebling.com
Supervision Alexis Georgacopoulos, Anniina Koivu, Thilo Alex Brunner and Luc Bergeron

Illustrations, graphic design and info graphics Helge Hjorth Bentsen
Illustrations Olli Hirvonen and Carolien Niebling
Photography Jonas Marguet
Photo collages and cover image Emile Barret
Making-of photography Noortje Knulst and Emile Barret
Mock-ups master butcher Herman ter Weele and molecular chef Gabriel Serero

Text Carolien Niebling
Rewriting Sarah Jane Moloney
Text editing Ariella Yedgar
Proofreading Sarah Quigley

Printing by ECAL/Benjamin Plantier
on a Heidelberg SX-52-5 press at ECAL, Switzerland
Cover printing by Ast & Fischer AG
Binder Schumacher AG
Paper Lessebo smooth natural 130 gr. from Fischer Papier AG
Font Tiempos, Acumin Pro

Editor
ECAL/University of Art and Design Lausanne
Avenue du Temple 5
CH-1020 Renens
www.ecal.ch

Publisher
Lars Müller Publishers GmbH
Pfingstweidstrasse 6
CH-8005 Zürich
www.lars-mueller-publishers.com

ISBN 978-3-03778-548-5

Acknowledgements
Alexis Georgacopoulos, Olli Hirvonen, Anniina Koivu, Thilo Alex Brunner, Luc Bergeron, Gabriel Serero, Herman ter Weele, Lars Müller, Marva Griffin, Benjamin Plantier, my family: Rob, Marianne, Maarten and Noortje Niebling, and friends, especially: Dominic Schlögel, Anthony Guex, Giulia Chéhab, Eleonora Castellarin and Ondřej Báchor. Thank you for all your help, advice, patience and loving support during the making of this book. Also a big thank you to all the kickstarter supporters who helped fund a part of the printing of the book.

Printed in Switzerland

éca l   **Lars Müller Publishers**